T0129660

Freedom and Evolution

Adrian Bejan

Freedom and Evolution

Hierarchy in Nature, Society and Science

Springer

Adrian Bejan
Duke University
Durham, NC, USA

ISBN 978-3-030-34011-7 ISBN 978-3-030-34009-4 (eBook)
https://doi.org/10.1007/978-3-030-34009-4

This Springer imprint is published by the registered company Springer Nature Switzerland AG
The registered company address is: Gewerbestrasse 11, 6330 Cham, Switzerland

Preface

Evolution is the defining phenomenon of nature. Everywhere we look what we see is evolving becaue it is free to move and morph. Without freedom to change, there is nothing, no design, no evolution, and therefore no future.

Freedom is everywhere because evolution (design change) is everywhere in the inanimate, animate and human realms. Yet, unlike evolution, freedom is not a scientific subject. This struck me as strange. It is as if scientists are afraid to speak the word "freedom" even though every day they rely on the reality (the physics) behind that word. For example, every single book of thermodynamics is filled with analyses and designs of "processes". The definition of process is the change in the description (the "state") of the system. Clearly, if the system is to change, then it must have in its description the property called freedom.

It's not surprising that freedom has been overlooked in physics. The more common a physical presence, the more likely it is that it is overlooked. It happened this way with gravity, sound, turbulence, and fish swimming. It took time for questioners to be born, and for physics to expand over the newly identified territory.

The purpose of this book is to present the predictive theory of evolution. It is to establish firmly the concepts of freedom and evolution in physics. The approach I have chosen is that freedom is physics, not opinion. This book is not about the politicians' narrative. All the ideas and examples in this book were published based on peer review in physics, biology, and engineering science journals. These sources are indicated for further consultation at the end of each chapter.

Like everything else that is physical (i.e., part of nature), freedom is measurable. Freedom is the measurement of how many physical features are free to be changed in the configuration of the system. Measurable is also the physical effect that the ability to change has on other measures of system performance such as efficiency, power, robustness, resilience, and life span. In human-made designs, freedom is also measured as the number of "degrees of freedom" that is present in the model (the facsimile) of a natural flow system. Degrees of freedom are those palpable features that can be changed freely and *independently* of other palpable features.

Along with freedom and evolution as physics, other concepts acquire a solid scientific footing: complexity, images, drawings, diversity, hierarchy, social organization, ideas, discipline, and the evolution of science itself. All these concepts belong in scientific discourse, in physics. Why physics, because physics is the science that covers everything. Its concepts have unambiguous meanings. They are useful, and they owe their existence to freedom as a physical feature of everything that moves, flows, and morphs.

Science evolves because of freedom, and freedom thrives because of science. It is easy to create in freedom—just think of the history of art and science. Look at where artists and scientists were and where they lived and created. Their names speak of geography, history, culture, wealth, and the physical movement of free people with ideas and freedom to question and change the status quo. Free migration was key to their salvation.

This book will empower the reader with a science that covers territories that so far were not associated with physics: economies of scale, diminishing returns, hierarchy, wealth, social organization, the spreading of ideas, and scientific thought. Empowerment happens in two ways: readers will understand better the world around them and they will apply that understanding to effect change faster and more effectively. Physics tells us why things must be the way they are, and also that they are the givens that you must know in order to improve life and society. Things that appear disconnected and random are fooling us: they are intimately connected, hierarchical, flowing together and along with us, and thriving because of freedom, organization, and evolution.

Society is an earth-size living organism. The larger city is a bigger and more efficient mover than the smaller. This is why when the city is thriving the smaller settlements and companies are joining the bigger, why people migrate from the countryside to the city, and why in the industrial age the global society is evolving from peasant to urban. When the city ceases to thrive, the migration is the other way, toward the countryside.

This story of science addresses head-on and nullifies contradictions that spring up in our minds, for example, freedom versus inequality, freedom versus rules and discipline, rigid hierarchy versus evolution, rules versus random diversity, and evolution versus seemingly stable design.

Key is the image of evolutionary design. Reading this book you will find yourself imagining "movies" of rivers flowing, animals running, pedestrians walking, and people riding on buses, trains, and airplanes. The fact is that nothing moves unless it is driven. Pushing comes from power, and power comes from fuel for machines and food for animals. Once a natural system moves, it continually evolves its configuration toward flowing more easily, a set of goalposts that progress keeps moving farther down the field. As systems evolve, grow, and become more efficient, they also become more complex. Why? Because joining and moving (flowing) *together* requires less power than moving individually. This is the physics basis of "economies of scale" and its most obvious manifestation: social organization.

The same physics principle accounts for the fact that river systems evolve into embroideries of small tributaries flowing into a major river, and why a peloton moves more quickly than an individual cyclist. The bigger stream, animal, and vehicle are more efficient movers than the smaller. The hierarchical system with many small and few large movers is more efficient than the "one size fits all", though also more complex.

It follows that in the life movement of a population, what we commonly refer to as the economy, the amount of fuel consumed by the population is directly proportional to its annual wealth, the gross domestic product (GDP). So, physical movement (the flow) and economics are two sides of the same coin. The same hierarchical flow architecture accounts for both.

I formulated several other "big questions", and I treated them the same way, based on physics. The complexity, diversity, and apparent unpredictability of nature are distilled in Chap. 1 to three main ideas:

First, designs are everywhere, around us and inside of us: tree-shaped flows, round cross sections, and rhythms such as inhaling and exhaling. Evolution of design is a universal, unifying phenomenon of nature, and it is predictable based on its own law of physics, the constructal law (p. 5).

Second, nature is a rich weave of "engines" connected to power-dissipating systems that act as "brakes". The engines and brakes move hand in glove, and evolve with freedom.

Third, humans and their contrivances (machines, artifacts, add-ons) are like everything else that moves and evolves on earth. None of the evolving nature would be possible without freedom.

It is easier and more efficient to move a unit of something together with other units (in bulk) than to move it alone. The big size helps in many ways, yet, not all moving things (rivers, runners, fliers) become one big mover. The reason is that on earth all the movement happens between a point and an area (or a volume), not along a line between just two points. On an area, movers have freedom to access in all imaginable directions. They "scan" the area with a hierarchy of movers, on patches that form a hierarchical mosaic. Economies of scale collide with the reality of space (areas, volumes), and give birth to hierarchy.

The diversity of hierarchical flow architectures covers the broadest spectrum accessible to human observation: all size scales, animate, inanimate, human made and not human made, and steady and time-dependent (Fig. 1). Examples detailed in this book are river basins, human settlements (city rankings), sizes versus numbers of trees in the forest, university rankings, and rankings of the highly cited authors.

The nonuniform distribution of wealth is predictable because it is due to the evolutionary architecture of all the streams of a live society. The physical movement on the surface of the earth evolves naturally as arborescent, hierarchical flows. The GDP of countries all over the globe is proportional to human movement because it is proportional to the annual consumption of fuel, which drives all human movement. More economic activity means more fuel consumption. Wealth inequality becomes more accentuated as the complexity of the movement becomes more accentuated. When added to the natural design, artificial design features

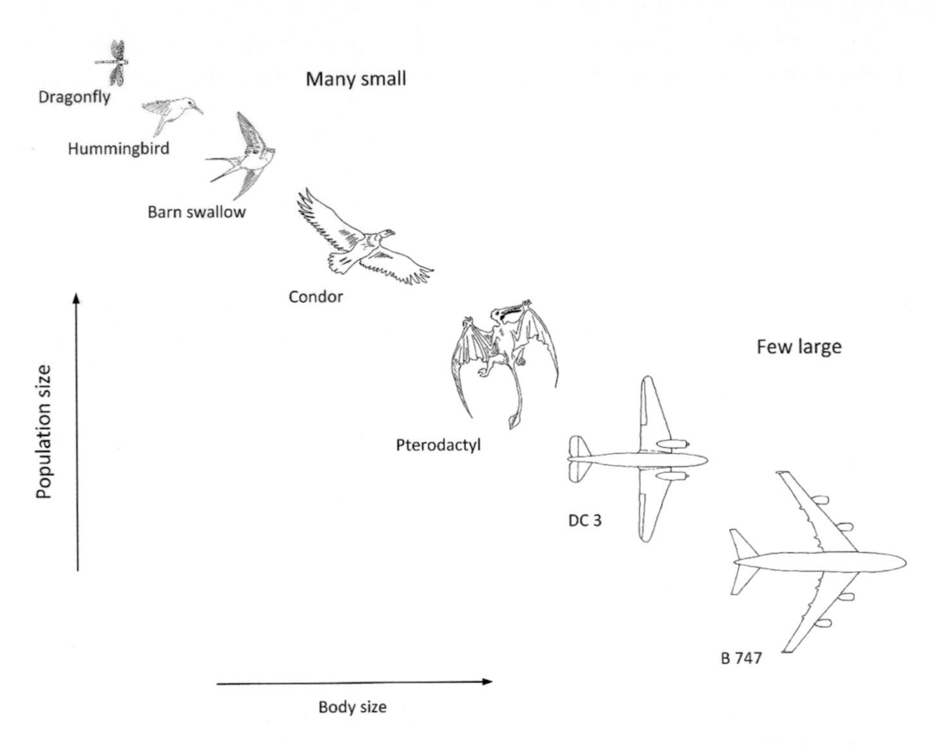

Fig. 1 Hierarchy in how animals and humans fly on the globe: qualitative distribution of population sizes versus body sizes (Drawing by Adrian Bejan)

decrease in wealth inequality. Trade routes are refracted paths for human movement on earth, which facilitate the movement.

Why does social organization happens by itself? Why does the organization becomes more hierarchical and unequal when the flows of society increase? In this book, such questions are answered from physics with two models, one without human society (river basins) and the other with human society (the distribution of hot water over an inhabited territory). The results from the two models are similar, except that in the model with human society the distribution of flow is less unequal. The reason is that the overarching presence of purpose (objective) at the scale of society controls inequality. Inequality persists even when the natural hierarchy of channels is replaced by artificial one-size channels everywhere. The spreading of individual innovation events over the populated territory benefits the whole society, in addition to benefiting the innovator.

There is no contradiction between freedom and reliance on discipline. In fact, the contrary is true: the scientist who possesses the disciplines is the most free to venture into new domains of knowledge. Disciplines are essential, empowering, and liberating for the scientist.

Complexity, organization, and evolution in nature are most powerful and useful when pursued as a discipline. A discipline has precise terms, rules, principles, and usefulness. A drawing has size, meaning (a message), and svelteness (the thinness of its lines). Drawings are useful when they are simple, easy to make, and not too large or too small. Icons and golden-ratio rectangles emerge from the human tendency to communicate more easily and faster. This is why features that are being perceived as attractive and beautiful are worth keeping.

Why is there so much diversity? Why don't we all look the same? Why are our occupations and contrivances diverse and becoming even more diverse? The answer lies in the freedom that all flow architectures have in how they spread, migrate, interbreed, and combine with flow designs that they encounter in their paths. This happened to the human flow on the globe. The human body architecture diversified as the first humans migrated out of Africa, to the north and the east. Along the way, humans have diversified as specimens of one "human and machine species".

The diversification of the machine part is evident in modern and contemporary times as the evolution of science and technology. The diversification of mechanics into thermodynamics and its many subfields accounts today for engineering sciences of many kinds: mechanical, civil, electrical, chemical, petroleum, nuclear, aeronautical, and more. The diversity of the human and machine species is of the same nature as the evolution of "niche construction" in animal evolution.

With freedom comes evolution with all its visible features: complexity, diversity, hierarchy, size, and free choices. Evolution—its future and its past—can be predicted. Detailed in this book are three predictions: one animate (animal locomotion), another inanimate (river basins), and the third about the human and machine species (aircraft). Many more predictions are available, for example, the cross sections of jets and plumes, the growth of snowflakes, the life span and life travel of animals and vehicles, the lung architecture, and the main measures of animal locomotion design (speed, frequency, force). Fundamental, for physics, is that a large architecture is not a magnified facsimile of a small architecture.

Finally, if evolution is so ongoing and everywhere, why do so many things look as if they are stuck in time? The reason is the phenomenon of diminishing returns, which is observed in freely evolving flow architectures that have become "mature". New changes have only marginal or imperceptible effect on the broad outlook and performance of the mature architecture. Examples that exhibit diminishing returns are the flows through tubes with freely morphing cross sections, vasculatures that connect the perimeter of a circle with its center, cantilever beams loaded with weight, and modern steam turbine power plants.

The evolution of science is a manifestation of the physics of freedom, access, and social organization. The physical movement of the individual generators of science is organized nonuniformly, hierarchically on the earth's surface. As society develops, it moves more, produces more, and generates more changes when it is endowed with freedom, free questioning, and self-correcting.

Freedom to change is the track on which the evolution train runs to join the other trains of science (e.g., Fig. 1.3). It is useful to know why the track is so "smart" that it led the designs of geophysics and biology to levels of perfection that continue to amaze us. It is useful to know how to use the evolution track so that our own artifacts evolve faster and more economically to even higher levels of efficiency, so that our own life as the human and machine species continues to become more free.

Give nature freedom, and nature comes back to life.

Durham, NC, USA Adrian Bejan

Acknowledgements

This book had a lot of help. I thank my family and my friends for supporting my work and my ability to continue during this writing project. I thank Deborah Fraze for typing and editing the manuscript and graphics, and for being close to me in my work since 1994. I thank my wife Mary for overseeing my work, its essence and public presentation, and for supporting me during my more daring moments.

I thank my closest collaborators who pioneered the physics of life, design and evolution everywhere: Sylvie Lorente, Marcelo Errera, Heitor Reis, Luiz Rocha, Antonio Miguel, Jordan Charles, Stephen Périn, Jose Vargas, Juan Ordonez, Giulio Lorenzini, Cesare Biserni and Pezhman Mardanpour. I am particularly grateful to my doctoral students Umit Gunes and Abdulrahman Almerbati, who made many of the figures in this book, and to George Tsatsaronis, Jose Lage and Mohamed Awad for their support on thermodynamics.

Good ideas bring interesting minds together, unexpectedly. From human events new ideas and better writing flows. I thank the following thinkers for taking an interest in my work and for teaching me how to think, speak and write better: Victor Niederhoffer, Peder Zane, Ephrat Livni, Michael Luby, Deborah Patton, Matthew Futterman, James Taranto, Malcolm Dean, David Troy and Anthony Kosner.

Acknowledgments

Contents

About the Author

Adrian Bejan received the Benjamin Franklin Medal (2018) and the Humboldt Research Award (2019) for thermodynamics and the consructal law of natural design and its evolution in science and social systems.

His degrees are from the Massachusetts Institute of Technology: B.S. (1971, Honors Course), M.S. (1972, Honors Course), and Ph.D. (1975). He was a Fellow in the Miller Institute for Basic Research in Science, at the University of California, Berkeley (1976–1978). At Duke University, he is the J. A. Jones Distinguished Professor since 1989. He has authored over 30 books and 650 peer-refereed journal articles, and has been awarded 18 honorary doctorates from universities in 11 countries, from France to Azerbaijan, and from Brazil to South Africa.

Professor Adrian Bejan's impact on thermal sciences is highlighted by original methods of theory, modeling, analysis and design that today are associated with his name: life and evolution as physics, constructal law, entropy generation minimization, scale analysis, heatlines, temperature-heat (T-Q) drawings, and many more. He has received the highest international awards for thermal sciences, and is a member of the Academy of Europe.

Other Books by Adrian Bejan

The Physics of Life, St. Martin's Press, 2016
Design in Nature, with J. P. Zane, Doubleday, 2012
Design with Constructal Theory, with S. Lorente, Wiley, 2008
Shape and Structure, from Engineering to Nature, Cambridge University Press, 2000
Advanced Engineering Thermodynamics, Fourth Edition, Wiley, 2016
Entropy Generation through Heat and Fluid Flow, Wiley, 1982
Entropy Generation Minimization, CRC Press, 1996
Thermal Design and Optimization, with G. Tsatsaronis and M. Moran, Wiley, 1996
Heat Transfer, Wiley, 1993
Convection Heat Transfer, Fourth Edition, Wiley, 2013
Convection in Porous Media, with D. A. Nield, Fifth Edition, Springer, 2017.

About the Author

Chapter 1
Nature and Power

People like to say that nature is complicated and becoming even more complicated. A lot has been said about diversity, complexity, unpredictability in nature, and more recently about the law of physics that accounts for all such observations. In this chapter, I distill this body of knowledge to just three ideas:

The first is that designs (images with meaning) are everywhere, around us and inside of us. Most obvious and best known are the tree-shaped designs, the arborescent flow structures of the river basins, human lungs, lightning, vascular tissue, urban traffic, snowflakes, river deltas, global air traffic, and vegetation (Fig. 1.1).

Many other images go unnoticed, as if taken for granted. One class is the round cross sections of ducts, and they cover the board from blood vessels, pulmonary airways, and earth worm galleries to the "pipes" carved by rainwater in wet soil and the hill slopes of the smallest rivulets of the river basin. Technologies of many kinds employ round ducts, and for a good reason: they offer greater access to what flows, greater than in the absence of round cross sections.

Less known are the rhythms of nature, the designs that represent organization in time, not in space. In most places, the flows that sweep areas and volumes flow in two distinct ways. In the river basin, the water first flows as seepage in the hill slopes (by diffusion, called Darcy flow), and later as streams in river channels. This combination is the physics of what others call "anomalous diffusion". The first way is slow and short distance, while the second is fast and long distance. Mysteriously, it seems, the water spends roughly the same time by flowing slowly (as seepage) and by flowing fast (as channel flow). The equality of times is the rhythm, and it is predictable from physics.

Oxygen reaches the lung volume thanks to the same design of two flow mechanisms. The short and slow is the diffusion across the vascular tissue of the alveoli. The long and fast is the flow through the pulmonary tubes. Diffusion and tube flow take the same time, which is the time of inhalation. Carbon dioxide is evacuated in the opposite direction, from a volume (the thorax) to a point (the nose). The same two-way combination facilitates the flow of carbon dioxide, first by diffusion across

A. Bejan, *Freedom and Evolution*,
https://doi.org/10.1007/978-3-030-34009-4_1

Fig. 1.1 Live and dead trees on Kapinga Island of the Busanga Plains, Zambia (Hot air balloon photo at sunrise: Adrian Bejan). Under this forest, the soil is a vast and tightly connected hierarchical vasculature of fungi that transports to the live trees the nutrients from the fallen trees, leaves, and fruit. The hierarchy of the tree society is visible above ground: few large thrive together with many small. Like a country, the tree society is held together by the ground, which is a live flow system vascularized with a hierarchy of diverse flows of water, nutrients, and animal life, constantly morphing in freedom

alveoli walls, and later by tube flow at larger dimensions. Diffusion time is the same as tube flow time. Even more intriguing is the fact that the point-volume flow (inhaling) takes the same time as the volume-point flow (exhaling). The flow direction (in and out, inhaling vs. exhaling) is not the idea, the rhythm is.

The same temporal design governs the flow of nutrients via blood circulation. Diffusion across the walls of the smallest blood vessels (the capillaries) is the short and slow way to flow. Stream flow along vessels larger than the capillaries is the long and fast way. The diffusion time is the same as the duct flow time. This is true for both directions of flow, in the arterial system (from lungs to whole body) and in the venous system (from body to lungs). In both directions, the flow is a design consisting of two tree flows, a volume-point tree connected to a point-volume tree where the point is unique (the heart), and the volume alternates between body and lungs. The time scale (the heartbeat) is the same for both directions of blood flow.

Rhythm and tree design govern the flow of water on land. The area-point flow of the river basin is followed by the point-area flow of the delta. The point in this image is one: the entrance to the delta. I grew up at such a point, on the Danube. Vegetation design is the coupling of two trees, and the base of the trunk is the connection. The root system is the tree that carries water from the wet ground (a volume) to the base of

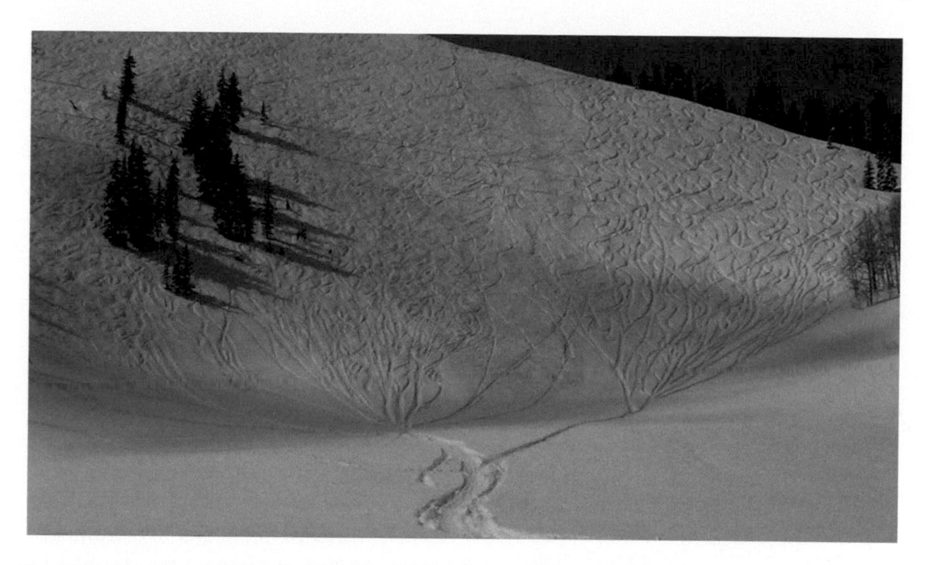

Fig. 1.2 How the ski slope "exhales" its population: hierarchical basin of skiers flowing down the side of a mountain covered with fresh snow powder (Photo: Rick Frothingham 2011; with permission)

the trunk (a point). The tree above ground carries the same stream of water upward, from the base of the trunk to the volume that contains the canopy and the dry air that blows through the canopy.

Diversity is part of the evolutionary design phenomenon. To appreciate its origin, think of how we all move through the seemingly rigid infrastructure of the city. We move with freedom. We make free choices all the time. Crowds flow as trees, from area to point and from point to area. In the early morning the crowd of commuters converges on the train station and the airport. It does so with hierarchy, with denser columns of people and automobiles on the larger and straighter streets that reach their point-size destinations. The canopy of this morning tree of human flow is the whole city area. In the evening, the same crowd flows in the opposite direction, from the point to the area (the city). In the morning, the city exhales people, and in the evening it inhales.

There are many other points (schools, offices, theaters, churches, and stadiums) that act as "valves" that open the sources and sinks for moving crowds. That is the design of human movement everywhere on earth, and it is like air respiration, inhaling (point to area) alternating with exhaling (area to point). It is a rhythm. Skiers on a snow-covered mountain illustrate how this flow architecture evolves (Fig. 1.2).

There are many more cases that indicate the time direction of design change, which we call *evolution*, and they seem even more dissimilar and unrelated in comparison with the examples mentioned above:

- Bigger flow architectures (river basins, lungs) are more complex, yet, their complexity is not changing; it is certainly not increasing in time, and not getting out of hand [1].
- All forms of animal locomotion (swim, run, flight) constitute a precise rhythm in which the frequency of body fluctuations (fishtailing, leg stride, wings flapping) is lower when the body is bigger. The speeds of bigger animals and aircraft are greater than those of smaller movers. Bigger movers can lift bigger weights, and have greater metabolic rates [2, 3].
- The bigger movers also exhibit longer life spans and longer distances traveled during lifetime. This holds true for all movers—animals, vehicles, rivers, winds, and oceanic currents [4].
- All flows that spread from one point to an area or volume (floods, snowflakes, human bodies, populations, plagues, science, inventions, political ideas) have territories that increase in time in S-curve fashion, slow–fast–slow. All flows that are collected from an area (or volume) have territories with S-curve histories, past and future. Examples are the evolving architectures of oil wells and mine galleries (tree-shaped, underground), each morphing and covering a territory that is expanding slow–fast–slow [5].
- Ancient pyramids and piles of firewood constructed by humans on all continents have the same shape. They are triangular when viewed in profile, with a shape that is as tall as it is wide at the base [6, 7].
- The earth's climate is stable, in a state of stasis with three distinct temperature zones, each with its own cellular currents and global circulation. This is the main design, with the earth as the intermediary node in the one-way flow of the solar heat current that the earth intercepts and then rejects to the cold sky. This design is predictable [8]. Superimposed on this design are small climate changes, also predictable [9], which are due to changes in the radiative properties of the atmospheric shell.
- Turbulent eddies (whirls) are born when the stream is thick and fast enough such that the rolling time of the eddy is shorter than the time needed by shear (viscous diffusion) to penetrate across the stream or the eddy. This rhythm is the phenomenon of turbulence, and it is why (contrary to established view) the phenomenon of turbulence is no longer an enigma since 1980 [10–12]. The origin of turbulence, which constitutes a constructal theory, should not be confused with computer simulations of highly complex (high Reynolds number) turbulent flows, such as weather prognosis, which are empirical, based on measurements and then modeling. Improving weather modeling should not be confused with predicting when and what turbulence *should* happen in a perfectly smooth (undisturbed) flow.

The list goes on, and so grows the impression that nature is complicated, diverse, unpredictable, and unruly. The examples cover the broadest realm imaginable: inanimate, animate, and human made. The impression is being proven wrong by the physics and biology literature that formed after the constructal law (1996):

For a finite size flow system (not infinitesimal, one particle, or subparticle) to persist in time (to live) it must *evolve* with *freedom* such that it provides easier and greater access to what flows.

In this statement, finite size means not infinitesimal, not one particle, and not one subparticle. It means a whole. Configuration (design, drawing) is macroscopic. Furthermore, to evolve and to persist in time as a flow system with configuration is the physics definition of to be alive. The opposite of that (nothing moves, nothing morphs, nothing changes) is the physics definition known as "dead state" in thermodynamics [13].

I wrote the constructal law when I knew a lot less about its implications. Now I see that the words "evolve, freedom, access" mean one thing. We all know what that thing is when we don't have it: freedom. That thing is what has been missing in physics.

All the dissimilar phenomena discussed above have been predicted by many authors by using the constructal law. So far, 12 Constructal-Law Conferences have been held all over the world, three sponsored by the U. S. National Science Foundation and one by The Franklin Institute.

So, that's the first idea. Evolution of design is a universal, unifying phenomenon of nature, and it is predictable based on its own law of physics.

The second idea is that all flow systems happen because they are driven by power. They flow and move because they are pushed, pulled, pumped, sucked, twisted, straightened, and shaped because of power. Power means work done per unit time on the system of interest. The work done is the product force $\times$ displacement, where the force is applied by the environment on the system and the displacement is the travel of the point of application of the force. The travel is relative to the frame of reference of the environment.

In sum, work entails movement, deformation, morphing, and change. The spent power is the cause of movement and change per unit time. The spent power is what drives the tape of evolution, the movie of design change over time.

The power comes from engines of all kinds and sizes. All engines happen naturally, the geophysical, the animal, and the human made. Most engines are not made by humans. Like the wheel, the engine is a natural flow architecture, not a human invention that does not exist outside the human sphere. The wheels of the biggest engines on earth are the currents of atmospheric and oceanic circulation. The wheels of a much smaller engine are under the mouse: two spokes (two legs) for each wheel, one wheel in front, and one in the rear.

The environment opposes the movement of the flow system. It resists being pushed out of the way. Consequently, the power that causes the movement is dissipated instantly into heat, which is transferred to the ambient. The relative movement between system and environment acts like the brake on a vehicle.

In thermodynamics, brakes are known more generally as purely dissipative systems. In the simplest model, a purely dissipative system (or a brake) is a closed system that receives work and rejects heat to the environment ("closed" means that mass flows do not cross the system–environment boundary; closed does not mean

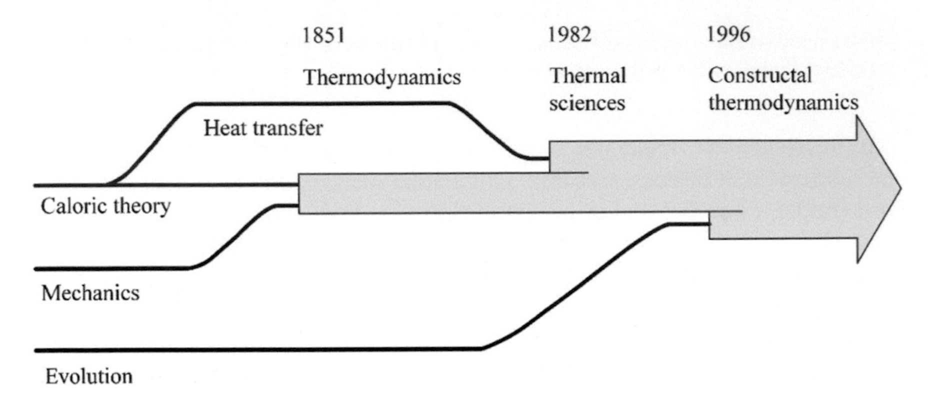

Fig. 1.3 The evolution and spreading of thermodynamics during the past two centuries

"isolated"). The system is in steady state, which means that its system properties (volume, pressure, temperature, energy, entropy) do not change in time. The brake system converts its work input fully into heat output. All brakes happen naturally; only a tiny number are made by humans, for vehicles.

Engines and brakes, all as natural phenomena, are the reason why design, evolution and constructal law emerged as a new domain and law in thermodynamics (Fig. 1.3) [13, 14]. Previously, thermodynamics was concerned solely with energy transfers (work, heat) between system and environment, and the conversion of heating into work and power, and vice versa. Classical analyses were based on the two laws that marked the merger of mechanics theory with caloric theory in the mid-1800s.

The first law states that energy is conserved. The difference between the energy flows entering the system (work transfer rate, heat transfer rate, enthalpy flow rate if the system is open) and the energy flows leaving the system is the rate at which energy is accumulated inside the system (Fig. 1.4, top). As such, the first law is the definition of energy, more precisely, the definition of the change in the energy inventory of the system, which is a system property that can be measured by invoking the law after having measured the heat transfer and the work transfer experienced by the system.

The second law is a concise summary of innumerable observations of the phenomenon of *irreversibility* in nature. Irreversibility means one-way flow, like the water over the dam, or under the bridge. Any stream (fluid, heat, rock avalanche) flows *by itself* in one direction, from high to low. Water flows in a pipe from high pressure to low pressure. Heat flows across an insulation from the high-temperature side to the low-temperature side. Rocks fall from high altitude to low altitude. Never the other way around.

The keywords in invoking the second law correctly are "irreversibility" and "by itself." Why, because a stream can be *forced* to flow the other way, from low to high. Water in the same pipe can flow from low pressure to high pressure if a pump is inserted between the low and the high, provided that power flows from the environment to the pump in order to push the water in the unnatural flow direction. Similarly,

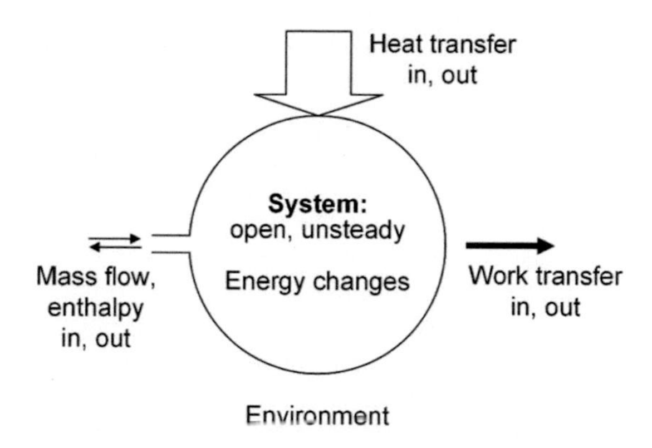

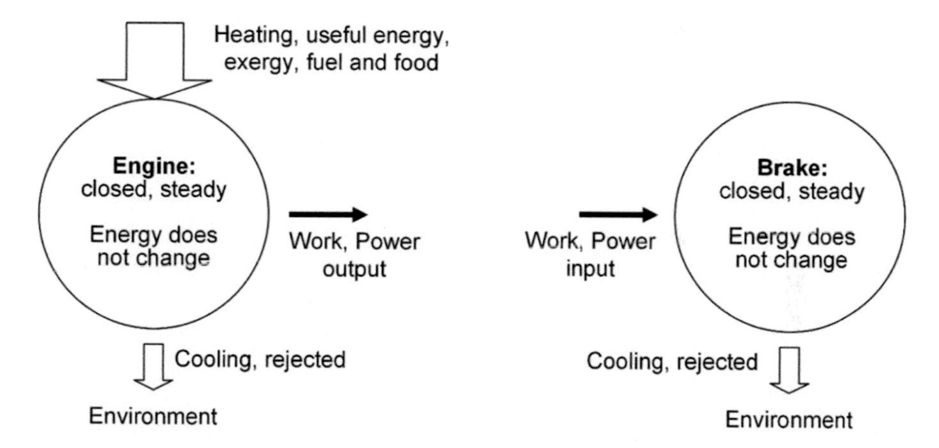

Fig. 1.4 Nature is in the eye of the beholder: it consists of just two systems, the system selected by the observer for analysis and discussion, and the rest, which is called the environment or the surroundings, also selected by the observer. Separating the two parts is the system boundary, which is chosen by the observer. Top: an open system can experience mass flow, heat transfer, and work transfer across its boundary. A closed system cannot experience mass flow, because its boundary is impermeable. Bottom: two classes of closed systems in steady state (not changing in time): engines and purely dissipative systems (brakes)

heat flows from low temperature (the cold zone) to high temperature (the room) in a refrigerator (the system), provided that power flows from the environment into the system in order to drive the compressor and "elevate" the heat current against its natural falling tendency.

The three founders of thermodynamics (Rankine, Clausius, Kelvin) cautioned against the subtleties of the new science. Here is the second law according to two of its original proponents in 1851–1852:

> No process is possible whose sole result is the transfer of heat from a body of lower temperature to a body of higher temperature (Clausius).

> Spontaneously, heat cannot flow from cold regions to hot regions without external work being performed on the system (Kelvin).

The second law is a most general statement, which says nothing about terms such as entropy, disorder, classical, and thermostatics. It is a statement of common sense, not jargon. It is not a mathematical formula. More recent reformulations of the second law in terms of disorder or entropy death are also correct but they hold for special and significantly narrower realms. The most general statement (from Clausius and Kelvin above) holds for "any system," and for the most universal manifestation of the second law phenomenon, which is the natural phenomenon of *irreversibility*.

Caution: Today, it is fashionable to assign revered labels (thermodynamics, entropy) to new concepts, and as a consequence, there are many "entropies" in circulation [15]. So, when you hear about the "thermodynamics" of this or that, ask that speaker to define his or her "system", and to show you the heating (*thermē*) and the power (*dynamis*) received or delivered by that particular system. Ask the speaker to define the "entropy" to which he or she refers. Fortunately, you do not have to wait forever for an answer, because the proper exposition of thermodynamics terminology and laws is widely available, for example, in Refs. [13, 14].

Entropy change (not "entropy") is a mathematical quantity invented by Clausius in order to express mathematically the irreversibility of *any* process (any change of state) experienced by any system. In entropy representation, the second law is an inequality, and the strength of the inequality sign is a measure of how irreversible (dissipative, lossy, imperfect) the process is. Here is a brief introduction to the terms in use:

The mathematical definition of entropy change is made with reference to a closed system that experiences a reversible process (from state 1 to state 2) during which it receives the infinitesimal heat transfer δQ [J] instantaneously from an environment of thermodynamic temperature T [K]. By convention, δQ is positive when entering the system, as in the functioning of a steam engine. The temperature of the system boundary spot crossed by δQ is equal to the same T, because during reversible heating or cooling there are no temperature differences and gradients anywhere. As the boundary temperature T changes during the process, so does the environment temperature. The definition of the change in the entropy of the system is the formula $S_2 - S_1 = \int_1^2 \left(\frac{\delta Q}{T}\right)_{\text{rev}}$, where S is a system property (a function of state, like volume and energy) and S_2 and S_1 are the inventories (the values) of entropy at states 1 and 2. The infinitesimal $\delta Q/T$ is the infinitesimal entropy transfer that accompanies the infinitesimal heat transfer δQ at the boundary point of temperature T. That is the new idea, heat transfer is accompanied by entropy transfer.

The second law must not be confused with the above definition, which is the definition of the property called entropy. The mathematical statement of the second law in terms of entropy is an inequality, and the inequality sign means "one way," or irreversibility. The simplest such statement is the one that holds for any closed system that executes any process 1–2, namely, $S_2 - S_1 \geq \int_1^2 \left(\frac{\delta Q}{T}\right)_{any}$. In this statement, T continues to represent the temperature of the boundary spot crossed by δQ, regardless of the presence or absence of temperature gradients.

One way to state the second law in words is to say that for any process executed by a closed system, the entropy change ($S_2 - S_1$, the change in the state function called entropy) cannot be smaller than the entropy transfer into the system, $\int_1^2 \left(\frac{\delta Q}{T}\right)_{any}$. More general versions of the second law inequality of the preceding paragraph hold for more general versions of system definition, for example, for open (not closed) systems operating in general, time-dependent fashion [14].

Comparing the two mathematical statements in the preceding three paragraphs we see the difference between a reversible process 1–2, and a general (called irreversible) process 1–2. It means that in the limit where the general process is reversible, the inequality sign $\geq$ reduces to the equal sign. This is why it is correct to view the inequality sign as a measure of the severity of irreversibility (or the departure from reversibility) of the process 1–2. Irreversibility means a flow that proceeds one way, from high to low. The inequality sign in the mathematical second law is a measure of the severity of the fall from high to low.

During the past 150 years, the mathematics based on the S formulation of the second law (and on the definition of entropy change $S_2 - S_1$) has become the discipline of thermodynamics that underpins virtually every single calculation, design, and technology that powered humanity in the modern and contemporary eras. None of this has anything to do with design, organization, freedom, and evolution in nature. The evolution phenomenon is distinct from the irreversibility phenomenon, which is why the constructal law is distinct from the second law (Fig. 1.3).

All processes in nature are irreversible. In the theoretical limit imagined by the original founder (Sadi Carnot 1824), the irreversibility is negligible, the inequality sign becomes the equal sign, the process is said to be *reversible*, and the second law is an equation, not an inequality. Unlike the second law, the first law is always an equation, regardless of the irreversibility or reversibility of the process. In this book, it is not necessary to use such mathematics. It is in fact advisable to tell the story without mathematics and, especially, without using the word entropy. Figure 1.4 is a graphic summary of the thermodynamics discipline of the first law and the second law. The system is any imaginable region in space or amount of matter. This is why it is drawn empty, like a black box. The two laws apply to any system that would reside inside the box. This utmost generality is also the limitation of the two laws. Why, because the systems of nature are not black boxes. They have configurations endowed with freedom to change into new configurations. This is why evolution has become part of thermodynamics (Fig. 1.3).

Nature is a rich tapestry of superimposed "engine + brake" systems, all flowing, moving, churning with freedom, morphing and evolving. This live tapestry converts into heat (rejected to the ambient) all the power produced by all the engines that drive all the movement. Because every engine converts its heat input only partially into power and rejects the difference as heat to the ambient, and because the power is itself dissipated into rejected heat, it follows that the heat input (originally, from the

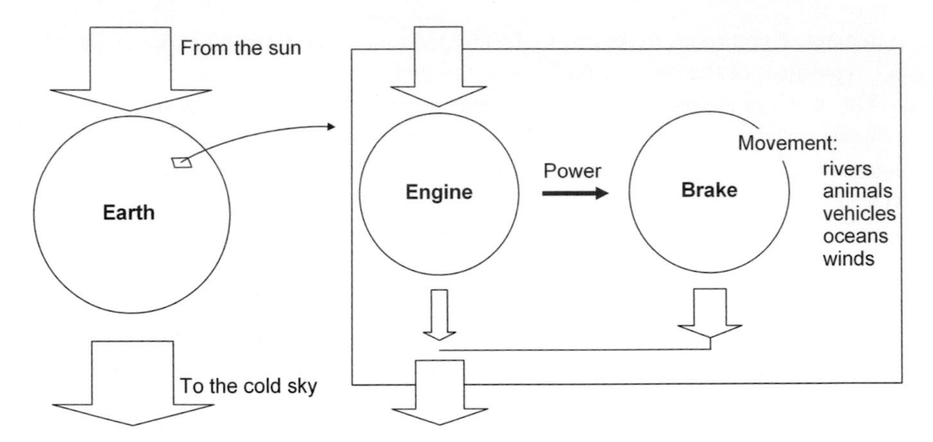

Fig. 1.5 The flowing tapestry of engines and brakes that mix and turn over the earth spheres, atmo, hydro, bio, and human

sun) to all the engines is rejected completely to the environment. For the earth-size engines mentioned previously, the environment is the cold sky.

Evolution enhances the mixing of the earth sphere. Inanimate and animate flow systems evolve in the same direction, to flow more easily and farther, by using and carving their environment, and carrying it with them.

The thermodynamics of the earth-size live flow system is very simple (Fig. 1.5). Heating from the sun drives the engine that the earth sphere is. The engine organs keep moving, flowing, turning, and morphing. The power generated by the engine is dissipated instantly while pushing all the things that flow and move against the friction and obstacles posed by their immediate surroundings. The wind, the vehicle, and the animal that move and displace the environment are engine + brake systems. The earth engine converts into power only a part of the solar heat current. The other part is rejected as a heat current to the cold sky, to the rest of the universe. The power produced by the earth engine (later dissipated by movement) becomes a second heat current that is also rejected to the cold sky (Fig. 1.5).

The bottom line is that the passing of the solar heat current "through earth" has the effect of constantly mixing the earth sphere, because of the running engine that drives the brakes that dissipate its power. Evolution is constantly improving the engine + brake design of earth so that the mixing is enhanced naturally, relentlessly. The live world is an immensely, highly diverse population of "vehicles," animal, geophysical, and human made, constantly moving and flowing as engine + brake systems. This tandem is the physics of what others called more recently "active matter" and "live matter".

So, that's the second idea. The sun heats the earth with a finite heat current, which is rejected 100 percent to the rest of the universe, like all the heat currents emanating from the sun. The earth is special, because it has properties that give birth to wheels of engines and brakes. Thanks to the freely evolving design of the engines + brakes tapestry, the sun drives the mixing and turning over of the earth's surface. This is why life happened on earth, and how it may be found on other planets.

The third idea is that humans are like everything else that moves and morphs on earth. We are driven by power from our engines (human body, metabolism and muscle power, domesticated animals, and vehicles), the engines consume fuel and food, and our movement dissipates the power completely. This flowing architecture evolves with freedom in a discernible time direction, which is captured by the constructal law on page 5. Our movement (our life) reshapes the earth in a progressively greater and more effective way over time, like the rivers and the animals.

Humans are not naked bodies. Each of us is much bigger and more powerful because of the contrivances (add ons, artifacts, the artificial) that are attached to us. We are encapsulated in our ingenuity, which is represented physically by our artifacts. We carry them with us, and they carry us with them.

Contrivances enhance the effect of human effort, from the shirt, knife and fork to the rope, pulley, and the automobile. The ancient Greek word for contrivance is *mihani*, which is the origin of modern concepts such as machine, mechanics, mechanical, mechano, mechanisms, and machinations. The word "machine" covers all the imaginable artifacts that empower humans. The "mechano" realm illustrates the meaning and direction of the constructal law, as summarized in the title of my 2000 book "Shape and Structure, from Engineering (i.e., mechano) to Nature" [2]. This is the physics behind new terms such as "mechanobiology" and "technobiology".

We are the "human and machine species". Each of us is a human and machine specimen that evolves during his or her lifetime. The machine part evolves every day through technology, commerce, science, language, writing, listening, recording, education, and artificial intelligence.

That's the third idea, and it emphasizes the pivotal role of all three. The physics of freedom and evolution is about us. Freedom is access to power. This is why this new science is absolutely essential.

References

1. A. Bejan, S. Lorente, A. F. Miguel and A. H. Reis, Constructal theory of distribution of river sizes, section 13.5 in A. Bejan, *Advanced Engineering Thermodynamics*, 3rd ed. Wiley, Hoboken, 2006
2. A. Bejan, *Shape and Structure, from Engineering to Nature* (Cambridge University Press, Cambridge, UK, 2000)
3. A. Bejan, J.H. Marden, Unifying constructal theory for scale effects in running, swimming and flying. J. Exp. Biol. **209**, 238–248 (2006)
4. A. Bejan, Why the bigger live longer and travel farther: animals, vehicles, rivers and the winds. Sci. Rep. **2**, 594 (2012)
5. A. Bejan, S. Lorente, The constructal law origin of the logistics S curve. J. Appl. Phys. **110**, 024901 (2011)
6. A. Bejan, Why people build fires shaped the same way, Sci. Rep. **5** (2015), article 11270
7. A. Bejan, S. Périn, Constructal theory of Egyptian pyramids and flow fossils in general, section 13.6, in *Advanced Engineering Thermodynamics*, ed. by A. Bejan, 3rd edn. (Wiley, Hoboken, 2006)

8. A.H. Reis, A. Bejan, Constructal theory of global circulation and climate. Int. J. Heat Mass Transf. **49**, 1857–1875 (2006)
9. M. Clausse, F. Meunier, A.H. Reis, A. Bejan, Climate change, in the framework of the constructal law. Int. J. Global Warming. **4**, 242–260 (2012)
10. A. Bejan, On the buckling property of inviscid jets and the origin of turbulence. Lett. Heat Mass Transf. **8**, 187–194 (1981)
11. A. Bejan, *Entropy Generation through Heat and Fluid Flow* (Wiley, New York, 1982)
12. A. Bejan, *Convection Heat Transfer*, 4th edn. (Wiley, Hoboken, 2013)
13. A. Bejan, *Advanced Engineering Thermodynamics*, 4th edn. (Wiley, Hoboken, 2016)
14. A. Bejan, Evolution in thermodynamics. Appl. Phys. Rev. **4**(1), 011305 (2017)
15. P.T. Landsberg, Entropies galore! Braz. J. Phys. **29**(1), 46–49 (1999)

Chapter 2
Economies of Scale

One idea from the preceding chapter is that nothing moves unless it is driven. The environment resists the movement. The body of the flier or swimmer experiences "drag". The skin of the fish and the airplane fuselage experiences "friction," and the impala and the truck experience drag and friction. This is why nothing moves unless it is driven, pushed, or pulled. This universal feature of nature is accompanied by another universal feature—economies of scale—which is the universal tendency toward acquiring new configurations that facilitate movement, provided that there is freedom to change.

Economies of scale is the common observation that is easier and more efficient to move a unit of something along with other units (in bulk) than to move it alone. All flow systems (with power, movement, and freedom to morph) tend to unite and organize for this reason. In every such system, the resulting organization is a hierarchy of few large and many small, which is the antithesis to the one-size design with identical movers, and without organization.

Economies of scale is easy to predict, demonstrate, and teach as a simple lesson of physics. When exiting from a crowded stadium or theater it is much easier to step in the space vacated by the person in front of you, rather than elbowing your way alone through the stagnant crowd. Conga lines through the impenetrable jungle are prefigurations of social organization.

Economies of scale is a physics phenomenon because it is universally present. It is evident in the biological, geophysical, human, and social realms. Its manifestation can be measured and evaluated in physical terms. To illustrate, I composed two simple problems, shown here in Fig. 2.1. They can be solved with minimal knowledge of fluid mechanics, or they could be tried in the laboratory by measuring the pressure drop along pipes with water flow and the drag force on bodies immersed in water channel flow.

A. Bejan, *Freedom and Evolution*,
https://doi.org/10.1007/978-3-030-34009-4_2

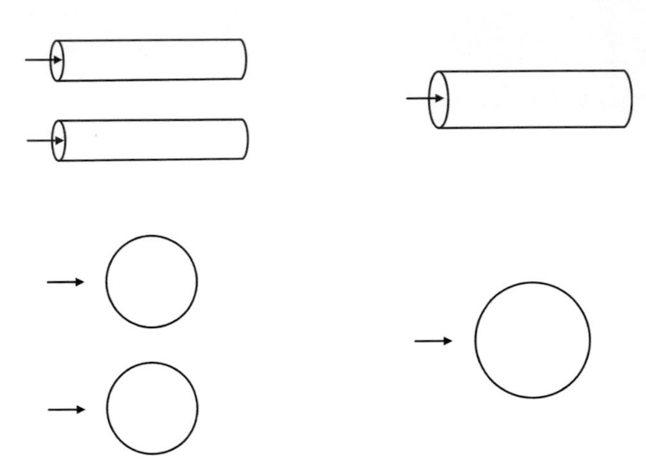

Fig. 2.1 Economies of scale happen when two or more flows come together and form one bigger flow. Water flowing through two identical pipes finds it easier to flow through one larger pipe. The total drag force on two bodies immersed in a stream decreases if the two bodies are replaced by one larger body. On a river, a tugboat that pulls two barges uses significantly less power if the two barges are replaced by one large barge loaded to the brim with the same amount of coal as the first two barges

Consider first the upper half of Fig. 2.1. There you see two identical pipes with identical flow rates of water, identical end-to-end pressure drops, and identical pumping power requirements for maintaining those flows. Will the pressure drop and the pumping power increase or decrease if the two water streams flow together in one larger pipe as long as the first two, and with a volume equal to the sum of the first two?

The answer is that the pressure drop will decrease, not increase. According to the constructal law, this means that when freedom to change is a physical characteristic of the flow system the design with two tubes should evolve toward the bigger tube, the wider channel. The decrease in pressure drop that results from this change is significant. If the flow regime is laminar and fully developed in all the tubes, small and large, the pressure drop decreases to half of its original value. If the flow regime is fully developed turbulent, and if the pipe walls are rough, the pressure drop decreases to 71% of its original value.

In the lower part of Fig. 2.1, the same problem is formulated for an "external flow," which is the counterpart of the "internal flow" configuration that each of the pipes represents. Two spherical bodies are held in a uniform flow of water in a channel or in a stream of air in a wind tunnel. The spheres are identical and so are the drag forces experienced by them. Will the sum of the two drag forces decrease if the two spheres are replaced by one larger sphere with the same volume as the smaller spheres put together?

If the total drag force decreases during this design change, then the power required for maintaining the water channel flow (and the airflow in the wind tunnel) decreases proportionally. The change in power requirement is sizeable, not imperceptible. If the surrounding flow is laminar and sufficiently slow (this regime is known as Stokes flow, or creeping flow), the total drag force drops to 63% of its original value. If the surrounding flow is sufficiently fast, with turbulent wakes and vortices shedding from each sphere, the total drag force decreases to 79% of its original value.

The things that flow in nature have the freedom to flow alone or together. The physics analyses presented above are accessible to everyone, and that knowledge is not new. New is the phenomenon that *free* objects that flow have the universal tendency to coalesce, to join, to organize, to get by more easily as they flow through their environment. This tendency is the new physics captured by the constructal law. It is for everyone to see in tea leaves that gather in the center of the bottom of a cup, after stirring. The same natural tendency is the cause of the Texas-size rotating clump of plastic garbage from China and India on the Pacific. It is the cause of all the animal locomotion in formations: flocks of birds, schools of fish, herds of migrating grazers, conga lines through the jungle, and pelotons of cyclists.

Birds do not have to fly to show us that joining makes life easier for them. Standing against the wind requires constant power expenditure in their leg muscles, as we will see at the end of this chapter. Facing a strong breeze on the roof, seagulls organize themselves in three ways (Fig. 2.2): they come together in a line, they face the approaching wind, and they stand behind each other at a distance of one body size, which is the scale of the wavelength of the buckled wake behind the preceding bird. In the elbow of the meandering wake, the backward rotating whirl slows down the wind, and the bird stands against the wind with less effort. Crane operators orient the cranes in a similar formation, in line, facing the wind, long tails in the rear.

The big size helps in many ways, not just in economies of fuel, power and cost during movement. One example is the peacock's tail, which until now was Darwin's

Fig. 2.2 Seagulls and cranes lined up and facing the strong morning breeze (Photos: Adrian Bejan)

unsolved puzzle. The unusually large tail and other strange features such as the big "eyes" colored on the plumage, are manifestations of the evolution of the animal to appear bigger than it is. A bigger animal is known for its proportionally bigger force, with which it can repel a predator or a competing suitor. Animals are aware of this. According to the constructal theory of animal locomotion [1, 2], the force with which an animal hits another is on an average equal to twice the weight of the animal.

Big size is key for the survival of the specimen. In the case of an animal that is not big (peacock, turkey, rooster, butterfly), the design features that helped maintain the flow of animal life on earth are those that made the animal seem big. The male turkey has not only a big tail that it can spread like a fan, but it also spreads and curves its big wings to the ground. When challenged, the male turkey looks as big as a wine barrel. For the same effect, the owl ruffles its feathers, and the hair stands up on the back of the angry cat and wolf.

When we marvel at these amazing features of body design, we are being descriptive, like most of the biology literature. What is the *cause* of these features? Why do they all empower the animal to look bigger than it is? Why is the unruffling of feathers and hair engrained in the behavior of the animal? The life of one animal is far too short for the animal to learn by trial and error what works as a deterrent in the face of danger.

The cause is always one, which is why it is both universal and a law of physics. It is the tendency of the flow of animal mass, in all its life forms, to configure and reconfigure to flow more easily, farther and with greater staying power, for greater access to its space on earth. People are no different, because they are also an evolving species—the human and machine species. Throughout history, nations and armies acted bigger than they were. Examples that come to mind are the Potemkin village, banging drums, fake cannons on the hillcrest, and cardboard tanks and airplanes on the ground. Physics are all such artifacts (they are "machines" in the most general sense), and physics is also the human tendency to acquire artifacts and to construct niche and social organization.

Wider channels and bigger bodies are more efficient movers. Larger surfaces are easier frontiers to trespass by heat currents and mass currents. The individual performance of channels, bodies, and surfaces is easy to predict based on simple analyses similar to those that address Fig. 2.1. In a large and complex flow architecture, the phenomenon of economies of scale is felt in the greater overall "efficiency" of the whole, when the whole is bigger. This is true across the board, from rivers and animals to vehicles [3] and social organization (Chap. 5).

The evidence that efficiency increases with size is massive. Compilations of efficiency versus size data for power plants of all kinds are shown in Refs. [4, 5]. Helicopter engines (Fig. 2.3) and steam turbine power plants (Fig. 2.4) illustrate the effect of size on efficiency. This universal trend has been predicted analytically not only for human-made movers [4–6] but also for movers not made by humans [3].

Animal movers have theoretical efficiencies (η) that increase in proportion with the body mass (M) raised to a power that is close to 1/2. This is similar to the motors of vehicles, which have efficiencies proportional to motor size (M) raised to a power

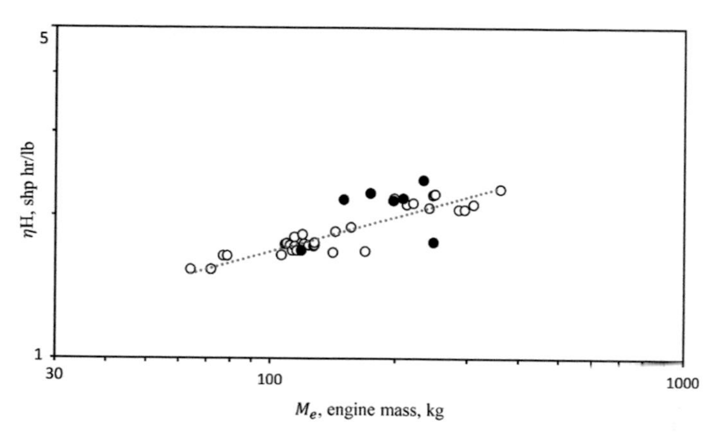

Fig. 2.3 Bigger helicopter engines are more efficient: the abscissa is the engine size and the ordinate is the work delivered by the engine per unit of fuel burnt [6]. The black dots are military helicopter data

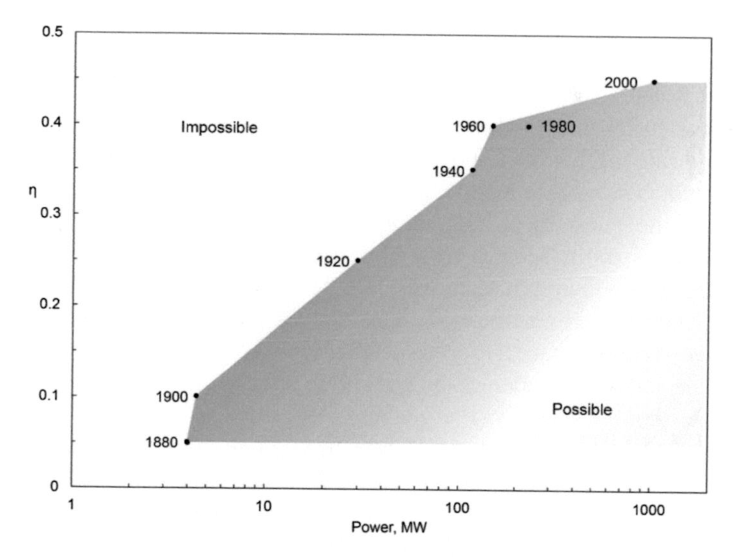

Fig. 2.4 The effect of size (power) on the efficiency (η) of steam turbine power plants during their entire history. The most efficient are the points that mark the crest of the mountain of designs, which appears shaded. The crest rises at a decreasing rate: this indicates diminishing returns from innovations that are being added to a mature technology (cf., Chap. 10). The efficiency η is the ratio of the power output divided by the rate of heat input to the steam that circulates through the power plant. This ratio is also known as the first law efficiency

less than 1 (e.g., Fig. 2.3) [6]. The rate of food consumption (or the metabolic rate) of an animal is proportional to the body mass (M) raised to a power α, where $\alpha = 2/3$ in the limit where the body is small and cooled by conduction, and $\alpha = 3/4$ in the limit where M is large and cooled by convection [1, 7]. This is worth contemplating, in view of the following exercise.

Two animals, each of size M_1, consume food in proportion to $2M_1^\alpha$, while a bigger animal of size $M_2 = 2M_1$ consumes food in proportion to M_2^α. The ratio between the consumption of the big and the combined consumption of the two small, $M_2^\alpha/(2M_1^\alpha)$, is $2^\alpha/2$, which is sensibly less than 1, for example, 0.7 if α is close to 1/2. This is the physics basis of economies of scale in animal design.

If joining and looking big are the natural tendency, then why don't all the moving things (rivers, runners, fliers) become one big mover that offers the greatest economies of scale per unit of what is being moved? The reason is that on earth all the movement is on areas and in volumes, not along a line, from one point to another point. Movement is on areas and volumes because movers have *freedom* to access in all imaginable directions. The movements are point-to-area and area-to-point flows, as in river deltas and drainage basins. In many other places on earth, the movement is point-to-volume and volume-to-point, as in inhaling and exhaling and many other physiological processes.

Economies of scale collide with the reality of space and geometry. A single big stream or a single big bird cannot sweep completely an area or a volume. Two big streams or two big birds cover an area or volume better than one, but they leave big spaces between them. These interstices offer access and attract even smaller streams and moving bodies in order for the whole area or volume to be swept by the movement. This is how hierarchies emerge, and why the next chapter examines more closely the hierarchy phenomenon of nature.

Looking back at the ground covered so far, the two physics lessons taught by Chaps. 1 and 2 are that nothing moves unless it is driven by power, and that when freedom to morph is present the movement evolves in time to offer greater and easier access to the available space. I end this chapter with an example that emphasizes both lessons, especially the freedom to morph and to choose the next flow configuration.

Let's question the statement that nothing moves unless it is driven. A counterexample that seems to contradict that statement is a ball on top of a round hill. The ball can begin rolling at any arbitrary time without any external force being applied, and without ever violating any of Newton's laws. The counterexample is real, but the absence of forces that drive the movement is mistaken. The ball on the hill begins rolling because it is driven by the horizontal component of the gravitational force, which is infinitesimally small at the start. No gravity, no rolling. No force, no motion. The counterexample confuses the ball's *freedom* to start moving to the right or left with motion itself.

There are many examples of the class to which the ball on the hill belongs. In mechanics, this class is called unstable equilibrium (keyword: equilibrium, which means no movement; equilibrium means two "equal weights" at the two ends of the balance). In thermodynamics, this is the class of closed systems in constrained equilibrium. Once again, equilibrium means no movement.

Another example is a closed system (a capsule) divided into two chambers by a flimsy partition. Nothing moves. Both chambers contain air, and the pressure in one chamber is higher than in the other. Nothing flows, and that's equilibrium, in this case constrained equilibrium.

The constraint is the flimsy partition. Nobody knows when and where it will crack and start leaking. The moment and location of the crack is the *freedom* in the physics of the crack configuration. In this example, freedom is infinite: the crack can have any size, shape, location, and time of occurrence. This particular kind of freedom is better known as chance, as in what heavy truck happens to roll down the street on which the experiment with the partitioned closed system is being observed. None of this supports the claim made in the counterexample. Why, because when the air starts flowing through the fresh crack, it does so because it is driven, from high pressure to low pressure, in accord with the second law.

Ironically, the argument with the ball that is balanced on a hill is even more mistaken, because to maintain the ball on top of the hill requires uninterrupted work, which must come into the system (the ball) from the environment. The juggler is hard at work when he or she keeps the ball at the top end of a pole. The ball is constantly moving sideways, no matter in which direction, and so it falls slightly and relentlessly. The adjustments that the juggler makes in the position of the stick amount to lifting the ball to its original level, constantly. The juggler is part of the environment that keeps transferring work to the closed system with unstable equilibrium inside (the ball).

If the reader never juggled, then here is the most familiar version of the same circus act. Any person who stands up is an inverted pendulum, like a ball at the top of a stick. The ball is the center of mass of the person and the stick is the distance from the center of mass to the feet on the ground. The person does not tip over because the incipient falling is counteracted and reversed instantly by the lifting forces provided by the feet, the heels, and the balls of the feet. The feet keep lifting the body to maintain it vertical and seemingly stationary.

This relentless work goes unnoticed because most people do not spend lots of time standing up to attention. To coerce a person to stand up straight is a form of torture. The same physics is why to deprive a prisoner of sleep is a form of torture. This is a Soviet scientific method of interrogation by torture, which was practiced by the secret police throughout the Soviet block. It was presented accurately by the German film *The Lives of Others* (2006).

References

1. A. Bejan, *Shape and Structure, from Engineering to Nature* (Cambridge University Press, Cambridge, UK, 2000)
2. A. Bejan, J.H. Marden, Unifying constructal theory for scale effects in running, swimming and flying. J. Exp. Biol. **209**, 238–248 (2006)
3. A. Bejan, Why the bigger live longer and travel farther: animals, vehicles, rivers and the winds. Nat. Sci. Rep. **2**(594). https://doi.org/10.1038/srep00594 (2012)

4. A. Bejan, A. Almerbati, S. Lorente, Economies of scale: the physics basis. J. Appl. Phys. **121** (2017), article 044907
5. A. Bejan, S. Lorente, B.S. Yilbas, A.Z. Sahin, The effect of size on efficiency: power plants and vascular designs. Int. J. Heat Mass Transf. **54**, 1475–1481 (2011)
6. R. Chen, C.Y. Wen, A. Bejan, S. Lorente, The evolution of helicopters. J. Appl. Phys. **120**, 014901 (2016)
7. A. Bejan, The tree of convective heat streams: its thermal insulation function and the predicted 3/4-power relation between body heat loss and body size. Int. J. Heat Mass Transf. **44**, 699–704 (2001)

Chapter 3
Hierarchy

Hierarchy is everywhere we see flow, movement, and vestiges of things that were once flowing. This phenomenon consists of few large and many small, flowing together, in stasis. When its physics origin is not questioned, hierarchy is often described as complexity, networks, turbulence, inequality, and diversity.

Hierarchy is the visible manifestation of freedom, economies of scale, and the configuration "choices" that flow systems seem to make to enhance the access to the finite space that is available to them. Hierarchical systems cover a hugely diverse territory in science. In this chapter, I sketch the backbone of that body of knowledge with just two strokes of the idea pen:

The first is that the diversity of hierarchical flow covers the broadest spectrum accessible to human observation: all size, scales, animate, inanimate, human made and not human made, and steady and time-dependent. Here are a few examples, and note that they come from overtly dissimilar domains of knowledge.

Under the falling rain, the ground surprises us with rivulets that arrange themselves into an all familiar "tree" configuration. The tree flows and morphs, freely. It is alive. It keeps rearranging itself to flow more easily, to evacuate the water faster down the slope. Scientists call this evolutionary flow architecture by many names such as diversity, multi-scale, dendritic, fractal, and many more. They should take courage and call it hierarchy.

The streams fill, bathe, and define a living whole, a live flow system that is continuous and thriving hand in glove with its environment. This natural tendency is obvious, undeniable, and it repeats itself. The streams arrange themselves hierarchically so that a larger stream flows because of its tributaries. The reverse is also true, as the flow of tributaries is possible only because the larger stream is draining them, bundling them and flowing as one. Harmony of movement is hierarchy, and it happens naturally. Hierarchy is good for the life and performance of the whole.

Hierarchy is a good word for expressing what we see, which is that a few large flow together with many small. Compared with hierarchy, the word complex, or complicated, does not say much because complex means twisted together. Inappropriate

A. Bejan, *Freedom and Evolution*,
https://doi.org/10.1007/978-3-030-34009-4_3

are also the words multi-scale and diversity, which suggest segments of many sizes in a stick drawing, or balls of many sizes dumped into a sack. Even Aristotle's line "the many and the few" does not capture the physics, because in reality the many are always small and moving slow and short, and the few are always big and moving fast and long.

We do not need to imagine rivers and river basins to see the natural origin of hierarchy. Players on a team know and use hierarchy every second. When I was a player, I observed that hierarchy happens as soon as the basketball coach puts a fresh group of players on the court. Every player knows it and uses it for the good of the team. After the first game, all the players know which player should receive the ball more often, because that player is a better shooter, and because this other player stands like a tower under the basket. Hierarchy is good for the access of the ball from the area (the court) to the point (the basket). Because of the same physics, hierarchy is good for the access of rainwater from the area (the river basin) to the point (the river mouth).

During my academic career, I observed that a newly formed committee organizes itself hierarchically the same way, naturally and spontaneously. After the first argument, every committee member knows who is the vision generator, the discussion leader, and with whom to associate to benefit from being on that committee. Advice to the revolutionary: if you have a different vision, do not join a committee, form your own.

In my own department at Duke, hierarchy and "diversity" emerged naturally, unnoticed. Those were the proverbial good old days. They were not the result of commands from above, although commands kept coming. They happened because of freedom in the pursuit of new ideas and forming and empowering the students' lives. We resembled a top sports team where origin, skin color, and passports did not count. The new player who knew how to play, played. Along with diversity came a hierarchy of talent and ability, which is good for the whole. This is the design of the great university of the good old days.

Without hierarchy, humanity would not have evolved to have language, religion, science, books, army, government, university, library shelves, and grocery shelves. The word hierarchy comes from the Greek word for chief priest, who by all accounts is a good person. Unfortunately, hierarchy acquired a negative interpretation following the French revolution, when the lack of social mobility and the nonuniform distribution of wealth in the old order (feudalism) was equated with inequality and immortalized in the slogan *Liberté, égalité, fraternité*!

The *égalité* was institutionalized overnight by similar revolutions, and, as if by magic, the next morning a new hierarchy emerged in place of the old. It happened this way with Wikipedia, which in the beginning was a wall on which any volunteer could affix new posters and correct old ones. Overnight, this writing activity was organized under an anonymous hierarchy of few editors and many trolls who write with "authority," delete what new volunteers write, and cite publications that surprisingly come from the same few sources. This is how we begin to guess who the anonymous members of that hierarchy are.

In the next chapter, we will see that nonuniformity in movement does not mean inequality. It means the absence of a single size in the flowing architecture that evolves naturally because it has freedom. The single size is absent because it is not part of nature. Look at the lung, the river basin, the city traffic, and the rest. They all have few large and many small streams flowing together, which is why they look the same even though one is animal design, another is geophysical flow architecture, and the third is the fabric of human social organization.

Given the *liberté*, the flowing whole equips itself with a hierarchical nonuniformity that enables every organ of the whole to flow (to live) as well and as easily and economically as possible. With liberty, each organ has *égalité*—equal access to change, to collaborate, to contribute, and to associate to pursue an easier, longer, and safer life.

Hierarchy is an integral part of the natural design of the flowing landscape and living world. The flows of nature evolve in time such that they flow more and more easily, for greater access. They attain this ever improving quality through the generation of flow design, that is, by acquiring configurations that evolve freely. Existing designs (drawings, literally) are replaced by new designs that flow more easily. In this mental viewing, we fit all the evolutionary scenarios of biology, the emergence of river basins and climate, and the evolution of technologies toward greater efficiency of human movement.

The flows that connect us as a society exhibit the same natural tendency to generate hierarchical flow configurations. Commerce and knowledge (science, education, news) flow in one direction: from those who have them to those who seek them because they are empowered by them. Those who receive them are set in motion, new territories open up for them, and they become freer and wealthier. When both ends of each such river basin have them and know them, the flow stops. What is not new does not travel.

Flowing leads to easier flowing. In this mental image reside the hierarchies that are visible in all the flow systems that cover the world map. These architectures form a multi-scale weave of point-area and area-point tree flows, all superimposed, all sustaining everything that flows and sweeps the earth. One example is the hierarchy of channel sizes and numbers in all the river basins that have been catalogued. With the constructal law, we deduced that the number of tributaries that feed a larger channel should be approximately four [1]. This prediction is in agreement with Horton's empirical correlation of river numbers, which states that the observed number of tributaries falls in the range between three and five [2].

Another obvious hierarchy is in the numbers of cities of the same size on large areas such as a continent (Fig. 3.1). The distribution turns out to be almost linear when plotting logarithmically the size of the settlement versus its rank. This line with slope in the range between $-1/2$ and -1 is known as a Zipf distribution, and it is found empirically (i.e., by observation) in virtually all the natural flow systems that connect discrete points with finite areas or volumes. The descending line in Fig. 3.1 has been predicted [3] by recognizing the flow access between two populations that live on each area construct (small and large) that covers the landscape. On every area construct (shown in white in Fig. 3.2), there are two groups that exchange flows: the

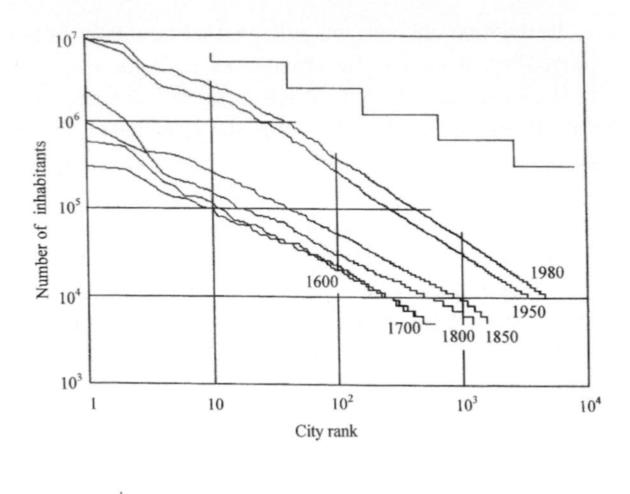

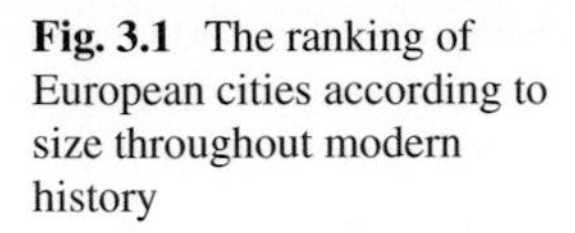

Fig. 3.1 The ranking of European cities according to size throughout modern history

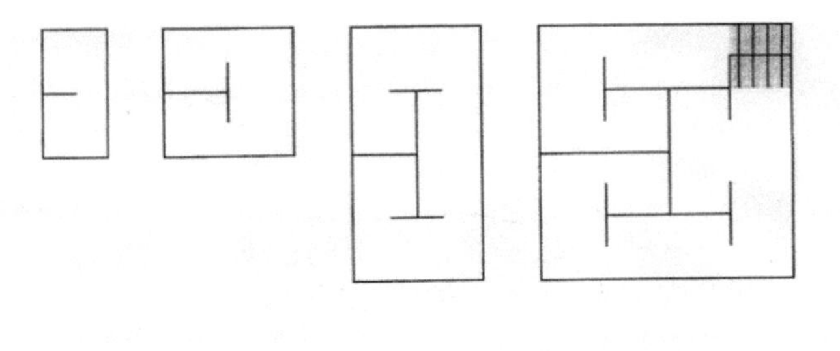

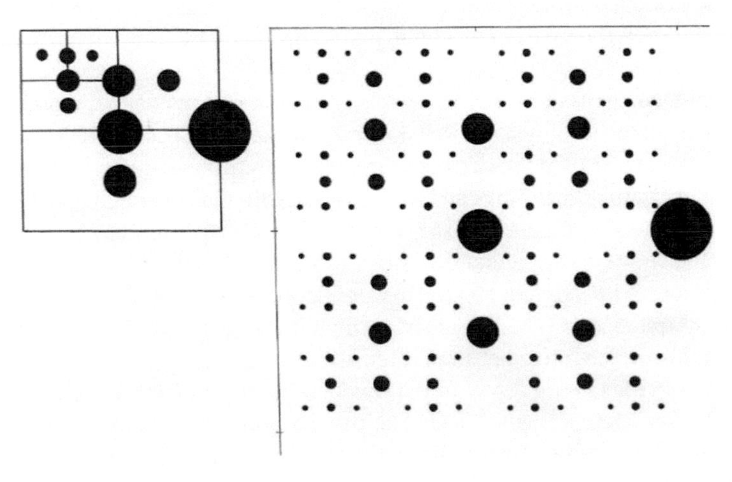

Fig. 3.2 The landscape of multi-rank cities and universities as a tapestry of hierarchically assembled areas

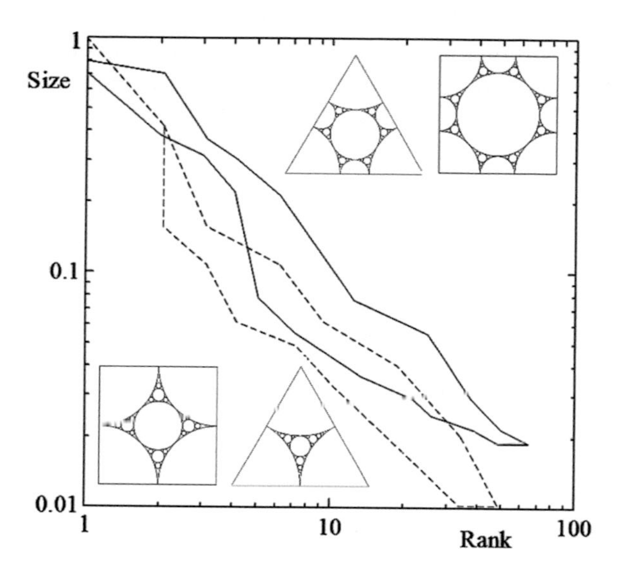

Fig. 3.3 Distribution of tree canopy sizes versus rank in the constructal design of the forest floor. The distribution is insensitive to the pattern (say, triangular vs. square) in which the multi-scale tree canopies are arranged on the forest floor. Pattern is neither design nor evolution

people who live on the land and those who live in a settlement (village, town, city), which is shown as a splat of black ink. We return to Fig. 3.2 later in this chapter.

Predicted is also the fact that the straight line will shift upward in time while remaining parallel to itself. This is a consequence of innovation, ideas, and technology evolution, which empower those who live on the land to achieve flow exchanges (e.g., production, trade) with larger and larger numbers of people living in the settlement. This too is in agreement with the history of the size-rank distribution over the past four centuries (cf., Fig. 3.1).

Hierarchical are also the sizes and numbers of trees in forest. In Fig. 3.3, the descending bands of size versus rank were deduced by arranging tree canopies of many sizes on the forest floor such that the entire floor facilitates the flow of water from the ground to the blowing wind. Two examples of arrangements (triangular and square) are shown in the upper right corner. The slope and intercept of the size-rank line are insensitive to the type of arrangement. Important is the hierarchy in the configuration of the multi-scale canopies that fill the forest floor so that the water flow rate upward, from the whole area, is facilitated. From this holistic view of free evolution came the prediction of the seemingly random and multiple scales of trees in the forest, and the alignment of the size versus rank data [4].

Society is a "live" flow system, perhaps the most complex and puzzling we know. It is the superposition of flow systems—a vast multi-scale system of flow systems—with organization (design), hierarchies, and time direction of morphing. It is a laboratory with evolution on display. It is the most difficult to comprehend because we, the minds that try to make sense of it, are small and deep inside the flow system.

Each of us is like an alveolus in the lung, an eddy in a turbulent river, or a vein on a leaf on a tree branch in the deep forest. From such a position of nothingness, which is identical in rank to the positions of enormous numbers of individuals, it is a formidable task to see and describe the big picture—the lung, the river basin, and the forest.

That's the first idea. Hierarchies are everywhere, including where it matters to us the most, in society. We will discuss social organization in greater depth in the next two chapters.

The second stroke of the idea pen is that while all hierarchies are happening naturally, not all are accessible to the human eye. Some are invisible by design or out of reach and not on our minds.

In society, hierarchies are of both kinds, those that are in plain view and known to everyone, and those that are invisible to most, and known only to a few [5]. In the field of social dynamics, the latter are known as dark networks and mafias. These are scientific concepts, not pejorative terms. An example of the first kind is the distribution of cities—their sizes and numbers—on a finite territory such as Europe (Fig. 3.1). In France, there is one big city the size of Paris, and many other human settlements that are obviously more numerous as they are smaller. This has been true since records are kept. Here is how to use flow evolution to predict the descending trend visible in Fig. 3.1:

Imagine an area element A_1 the inhabitants of which produce many streams (students, agricultural products, timber, game, minerals, and so on). The flow rates of such streams are proportional to A_1. These flow rates sustain a human concentration (a dot) located on A_1, where the number of inhabitants is N_1 and the production streams are of a different sort (education, knowledge, services, devices). There is an equilibrium between what flows from the area A_1 to the human concentration N_1, and what flows from N_1 to A_1. Key is that both classes of flow rates (area-point and point-area) are proportional to A_1, and this means that the size of the human concentration N_1 must be proportional to the size of the area A_1, which is allocated to the concentrated settlement.

The flows from the human settlement N_1 to the individuals spread over A_1 consist of goods and knowledgeable individuals, books, and science. The human concentration in this case is the city or the university campus [6], and the area A_1 is the territory that the concentrated human settlement serves. The constellation of cities on the landscape is a reflection of the area constructs of land-city counterflows that sweep the globe.

The distribution of human movement on the earth's surface is a natural construction of compounding area constructs, as shown at the top of Fig. 3.2. Like an area element in a river basin, which feeds the big stream that is discharged from the area, each area construct sustains the flows that reach a human concentration on the boundary of the construct. It follows that the human concentration on the boundary is proportional to the area size of the construct. If the human concentration is a university, then the size of the university (the generated flow of new ideas) is proportional to the size of the flow area that it serves. The landscape is covered by settlements ranked hierarchically because the area constructs have multiple sizes and are assembled hierarchically. The analogy between societal flow and the river basin is explored further in the next chapter.

Academia, like the population on a territory, is a laboratory to study the evolution of flow hierarchy in our lifetime. A few years ago, I documented the coexistence of the two kinds of flow hierarchies in society, the visible and the hidden, and for this I used the society I know best, which is academia [5]. For a hierarchy that is naked on the table, I used the world rankings according to the citations of scientific publications. For a hidden hierarchy, I used the rankings according to numbers of members of a national academy.

What the constructal law predicted for multi-scale river basins, demography, and forests also applies to the design of human flow on the same globe. Science and education flow naturally through a vascular body of student and professor paths to universities. Each university is connected to the entire globe, sustained, fed, and drained by the entire globe.

The older universities have dug the first channels, which are now the larger, immutable channels that irrigate the populated landscape. "Larger" does not mean a greater number of bodies moving in and out of classrooms. Larger are the streams of the more creative. The creative are the big channels that attract the special individuals who have the gift—the calling—to generate new ideas, and who develop disciples who produce and carry new ideas farther on the globe and into the future [7]. The swelling student population is served well by the memory built into the education flow architecture of the globe.

Out of one spring, say, Harvard, come lots of drops that fall on fertile ground. The history of the university system in America is that the drops from Harvard gave birth to Yale, and then there were two springs that gave birth to Princeton, Penn, and Columbia, all in a perfectly straight line on the map, from Boston to Philadelphia. That line is now the trunk of the big tree of university education in the U.S., which is the envy of the world. On that trunk, new and more specialized grafts took life, for example, universities with engineering and medicine orientations.

True is that in every house where the family is old, there are rules that have become rigid. Yet, if the family teaches the culture of free questioning, out of that home also emerges a stream of new thinkers who question the world and the status quo. That new stream wets the plain. This is the good news. It is what I have been observing, for example, in the growing immensity of publications these days. The highly cited authors are from everywhere, but a lot of them were educated on the trunk of the university flow system that holds, wets, and nourishes the world.

From this theoretical view followed the prediction that the hierarchy of universities should not change in significant ways [6]. The same rigidity and predictability characterize the hierarchy of the universities that win basketball championships [8]. This kind of hierarchy is as permanent as the hierarchy of channels in a river basin. It is in the open, above board. It is natural because it is demanded by the entire flow system—the globe—on which huge numbers of individuals pursue the same thing: physical changes in their lives, for easier and freer movement and living. The implementation of changes is what knowledge is [9].

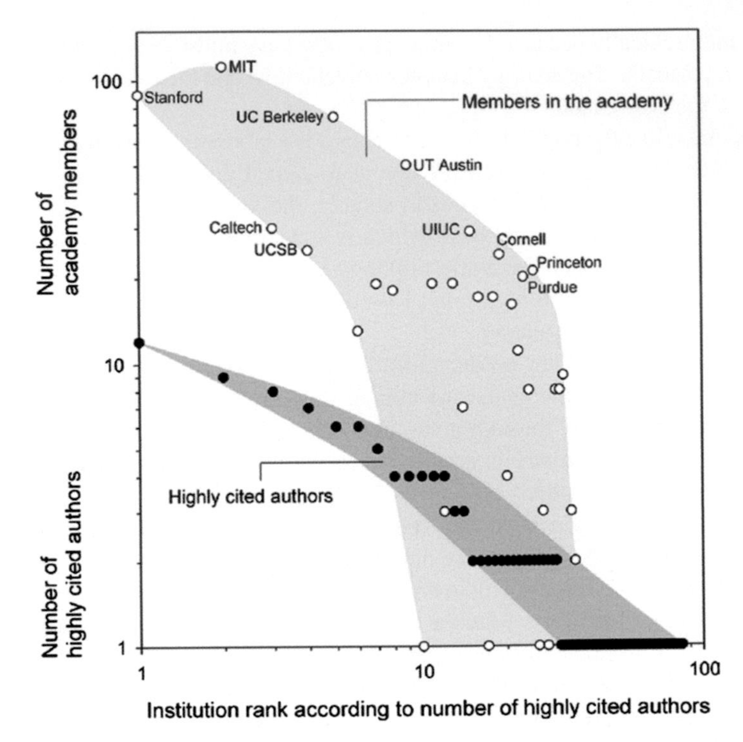

Fig. 3.4 In this figure, there are two descending bands of data because each university is represented by two points, which are aligned on the vertical. The abscissa shows the ranking of universities according to their number of highly cited authors: that number is indicated on the lower half of the ordinate scale. For example, the ranking begins in this order: Stanford, MIT, CalTech, etc. The upper half of the ordinate scale shows the number of academy members of each institution. The two bands of data are strikingly dissimilar

The rigid ranking of universities is a reflection of the reality that university fame is essentially the same hierarchy as the hierarchy of the individuals who generate the ideas for which the universities are known. The highly cited authors have a natural hierarchy because theirs is the result of the efforts and choices of very large numbers of researchers and practitioners who do not talk to each other before deciding to cite their sources. Even greater numbers appropriate and use the good ideas and do not cite their sources (cf., Chap. 11). The hierarchy of the most cited authors is an indicator of the nodes and areas (elemental constructs) of the flow of ideas. The highly cited hierarchy is shown in the lower part of Fig. 3.4, where the nearly straight band represents the ranking of universities according to the number of authors that each university has on the highly cited list. Important to note is that the lower band of data is nearly straight and has the same slope as what we saw in Figs. 3.1 and 3.3.

All flow structures are evolving and improving, yet some are hidden from view. The hidden are the channels where participation is based on personal connections, on who you know, and on who needs you for the safety and perpetuation of the dark network. Here is the difference between the clarity of the flow of ideas (Fig. 3.4, lower band) and the opacity of dark networks that permeate the same flow space. I took this example from engineering publishing [5], because this is the field I know best. Scientists from other fields can construct analogous examples by examining their own hierarchies of idea production and access to their national academy. Furthermore, the numbers fluctuate slightly from year to year, but the two patterns that emerge are so dissimilar that fluctuations do not matter.

The list of highly cited authors in the entire "engineering" domain (all disciplines, all countries, living and deceased, in February 2009) contained 253 names worldwide. The membership in the National Academy of Engineering (US alone) contained 2440 names. The 1:10 ratio between the two lists means that most of the members of the academy are not highly cited.

The contrast is even sharper. I placed this comparison on a common basis by removing from the 253 highly cited the 80 names of researchers who work in countries other than the US. I also removed the two names of highly cited authors whom I knew to be deceased (the academy deletes immediately the names of its deceased members). After these subtractions, which are conservative, the highly cited list shrank to 171 names. From the list of academicians, I removed the 197 who work in foreign institutions, and I kept the remaining 2243. The resulting comparison is between 171 highly cited and 2243 academicians, which form the ratio 1:13. Furthermore, only one-third of the highly cited authors (namely, 60) are also in the national academy, and they represent a shockingly small 2.7% of the 2243 members of the academy.

The contrast between the two bands in Fig. 3.4 is a reflexion of the fact that unlike on the highly cited list, which lies naked on the table, the path to the national academy is not visible. The vast numbers of researchers who vote for one highly cited author every day by using and citing ideas in the literature are not inside the academy, they cannot nominate, and they cannot vote for an idea generator.

The applicability of these findings is general. All steps of promotion, honors, and peer review in the science profession, all the way to the difficulty to publish in *Nature* and *Science*, can be analyzed in the way that the flow into the national academy was unveiled here. Peer review is not the culprit: the peer review system crumbled two decades ago because of the deluge of electronic publishing and the opacity that comes with that. It crumbled along with the old-fashioned ways of publishing, which meant respecting and protecting the original author's ownership of the original idea (cf., Chap. 11).

In complete accord with the constructal law, when one hierarchical flow structure crumbles it is replaced by another. Always. That's evolution. Instead of honest reviewing, today we see a global competition for citations and clever ways to game the citations system [10–17], which in an increasing number of cases is driven by nationalism encouraged by the government [18, 19]. This has led to the natural formation of forests of dark networks: the citations cartels [19–21] of invisible groups of authors (not co-authors) who cite each other's articles at every turn, gratuitously.

In the face of this physics, the challenge is to protect the precious and noble features of the western tradition of science, and to defend fiercely the merit system and the paternity of ideas. What is to be done then? Nothing, except a little advice to the young: the aspiring scientist should make a choice early in his or her career, and stick with it. If the scientist is creative, then the "highly cited" is the stadium in which to play. Take a hint from Ludwig Boltzmann, who wrote "I see myself better at integrals than at intrigue" ("ich verstehe mich aufs integrieren besser, als aufs intrigieren," Letter to Henriette von Aigentler, 1876).

In summary, the hierarchy in the knowledge "industry" today is a manifestation of the universal urge to live a better and easier life, which in modern society is measured as wealth: money or fuel spent with purpose is the physical measure of how good life is, while being lived. The hierarchy of universities is good for the lives and wealth of the professors and students who flock to the best universities. This is natural, and is why the university hierarchy happens, why it is rigid, not changing much. Hidden in the university hierarchy are the dark networks of researchers who pursue similar topics: science here, medicine there, engineering somewhere else. These are the national academies, known by those who are on the inside, good for those who are inside.

Hierarchy in academia flows the same as in a team sport. On the soccer field and the basketball court, the best are few and the near best are many. In sports, as in academia, the "near best" commit more of the fouls and more of the cheating. Some are very good at masking it, and are rewarded for it even though fouling, cheating, and plagiarizing threaten the life and career of the victim. In academia, the many who believe they should also be in the elite are the ones who cheat more and also write more of the vicious anonymous reviews of their peers' manuscripts. They come out of the closet when one of them, a critic, writes a book review that savages a truly original work.

> "Art critic! Is that a profession? When I think we are stupid enough, we painters, to solicit those people's compliments and to put ourselves into their hands! What shame!
> Should we even accept that they talk about our work?" (Edgar Degas).

So, that was the second idea. Hierarchies happen if the flow system is morphing with freedom over its area or volume, but many hierarchies are so big, distant, and obscure that they do not cross our minds.

Recent progress on the physics basis of hierarchy in nature continues to bring together phenomena that were previously not noticed. To the well-known animate and inanimate examples (animal locomotion, river basins, turbulence) covered by the constructal law, we now see examples that belonged traditionally to solid mechanics. The natural occurrence of hexagonal basalt columns is attributed to a principle of maximum "energy" release [22]. The occurrence of cracks in solids is based on the same principle [23–25]. Soil cracking under the drying wind is the constructal phenomenon of evolutionary design that enhances mass flow and accelerates drying [26]. The aggregation of dust particles into clusters and dendrites was shown to be the result of the same principle, and its effect is to relieve electrostatic forces of attraction faster, through the evolutionary design of configuration [27].

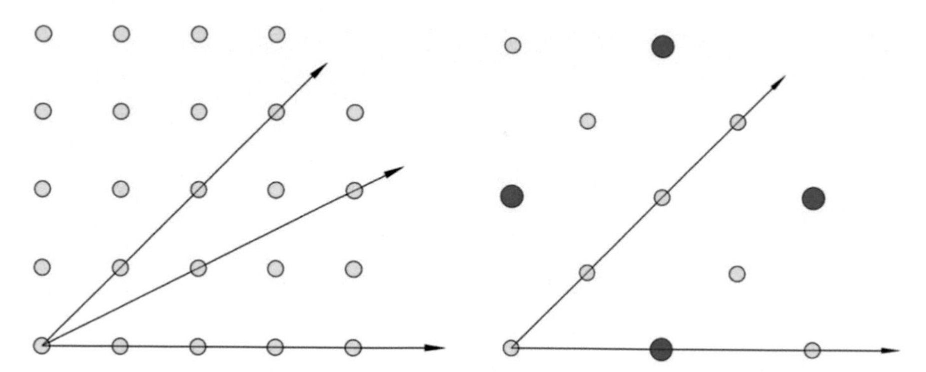

Fig. 3.5 Bodies of equal size (left side) are suspended equidistantly in a plane in outer space, and constitute a system with uniformly distributed internal tension. Only the quadrant of the system with the center in the lower left corner is shown. The forces on the body in the corner are the resultants of all the forces integrated along the radial directions viewed from that body. The bodies attract each other, larger bodies emerge, and the internal tension of the system decreases. This takes time. If bodies coalesce nonuniformly, with few large interspaced with many small (as on the right side), the internal tension decreases faster than if the original system (left side) is followed by bigger bodies of one size suspended equidistantly

Added to this growing list of evolutionary hierarchies is the example of bodies suspended in outer space [28]. Hierarchy emerges in two ways, through accretion (coalescence) and fragmentation resulting from collisions. Viewed from thermodynamics, the bodies in space form a system in a state of internal tension because of gravitational attraction between neighboring bodies. This system evolves freely by moving and changing its internal configuration. Bodies coalesce into larger bodies, and their collision (with fragmentation) dissipates the tension and the resulting kinetic energy, en route to reduced body–body attraction throughout the system. This phenomenon has been studied in celestial mechanics under several scenarios and is recognized as the basis for the formation process of planets and the asteroid belt [29].

Sizes increase over all scales through accretion. Yet, the natural phenomenon is not only the growth of the body sizes but also the emerging hierarchy. The fundamental question is why hierarchy happens in the first place, and why a uniform distribution of bodies of the same (growing) size must not happen. What causes the hierarchy? The gravitational effect alone does not explain the hierarchy of sizes of bodies in space. The additional effect is the natural evolution of the whole flow configuration during accretion, such that the flow and evolution of the system toward less tension are facilitated.

Consider a system of identical masses suspended uniformly in space (Fig. 3.5). Forces of mutual attraction keep the system in a state of internal tension. In time, the tension is relieved through the creation of aggregates. Two bodies attract each other with a force that is proportional to the product of the two masses and inversely proportional to their mutual separation distance squared. The shape and relative motion of the bodies are not considered. Assume that a space is filled initially with masses of one size (m) that are motionless and distributed uniformly. The spacing

Fig. 3.6 Drops of fish oil on water illustrate the phenomenon of hierarchical coalescence. Because of surface tension, the uppermost water layer is a two-dimensional system in tension. The freedom of the system is in its configuration of oil drops on water. The oil drops coalesce, the tension decreases, and hierarchy is the result. The hierarchy becomes more accentuated as time passes. The sequence of three photographs lasted approximately 20 min

between two neighboring masses (r) is the same everywhere. This suspension is in a state of uniform volumetric and isotropic tension.

The coalescence of small masses into larger masses is driven by internal tension, which becomes smaller as a result. No movement (i.e., death) would be characterized by zero tension and complete coalescence, with all the individual bodies collapsed into one large body. In thermodynamic terms, the system is isolated and exhibits internal changes (mass flows) that take it from an initial state of internal tension toward a final state of no tension and no movement.

The phenomenon is not the death of the system. Nobody will be around to observe that. The phenomenon is the life of the system en route to that end. Should the system evolve as equidistant masses of a single size that increases through coalescence, or should the evolving design be hierarchical, heterogeneous, with few large masses and many small masses that feed the large masses? With high school mathematics, we showed [28] that masses in a suspension have greater access to their neighbors and coalesce faster when they coalesce nonuniformly, hierarchically (cf., Fig. 3.5, right).

It is the hierarchy that is natural, not the uniformity. This is in accord with the physics principle and all observations of freely evolving flow systems everywhere. The natural tendency toward nonuniform coalescence can be visualized in the kitchen (Fig. 3.6).

Hierarchy also appears when compression is induced by a force at a point in a volume filled with a large number of granules that have freedom to shift and come in contact when compressed. This happens when a projectile impacts soil or sand. The volume is a flow system with untold freedom to morph, because the particles are free to rearrange themselves in practically an infinite number of configurations. Yet, the configuration that is born at impact is a tree of particles that are compressed the most, as if stuck together in solid columns, and their branches spread the point-impact force through the volume (Fig. 3.7).

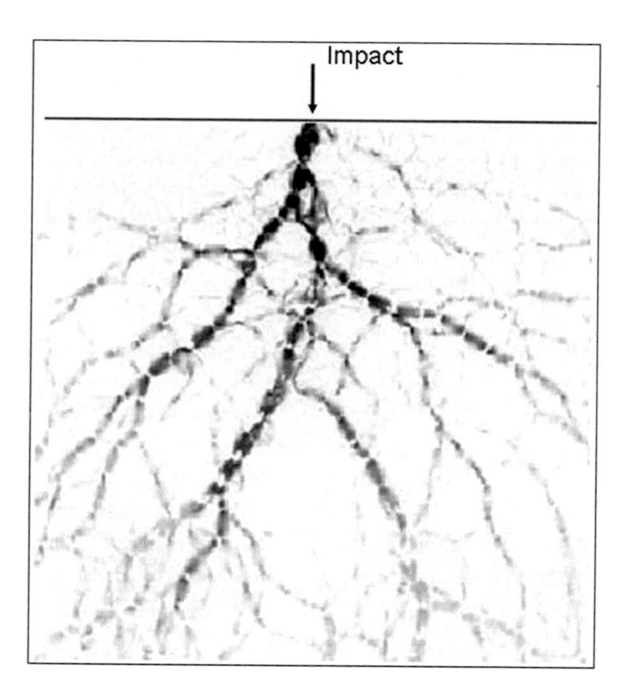

Fig. 3.7 The natural emergence of bones, skeletons, tree roots, and other solid members. When a force is applied suddenly on living tissue or soil, momentum is transmitted from the point to an entire volume by a spontaneous tree-shaped alignment of grains that transmit high stresses. The living system is a flow system for the flow of stresses. Its constructal-law tendency is to allocate mechanical strength (stronger and more material) along the channels with high stresses. These reinforcements become bones, tendons, skeletons, tree roots, and branches of roots (photograph courtesy of Prof. R. P. Behringer, Department of Physics, Duke University)

Earlier, I made the connection between this hierarchical alignment of solid and the origin of bones and skeleton [30, 31]. More recently, it occurred to me that the volume subjected to point impact is filled suddenly not only by a hierarchy of solid "bones," but also by a hierarchy of forces.

The compressive forces along the trunk and big branches in Fig. 3.7 are bigger than along the smaller branches. Under impact, the volume fills with forces that are distributed hierarchically. The forces *should be* distributed this way because of the same physics principle as in Figs. 3.5 and 3.6. With hierarchy, the compressed volume reaches its new equilibrium faster.

Figure 3.7 is the clue to the puzzle regarding the mother force that serves as common origin of the hierarchy of multi-scale forces that fill the universe. The tree of forces that connects the point of impact with the volume is the answer to one of the darkest mysteries in physics. According to the Stanford physicist Helen Quinn [32, 33], there are four fundamental forces in the universe: gravity, which you can feel pulling you down into your chair, the electromagnetic force, which binds the atoms in your chair, the strong force, which holds the atom's nucleus together, and the weak force, which is responsible for radioactive decay. Physicists have been trying to unify the forces for years.

The new idea is that the hierarchy of forces *should* have happened, and that the multi-scale hierarchy of forces should persist long after the beginning of creation. Just like the granules compacted in the canopy of its own tree of forces.

Another predictive idea that follows from this new connection is that the volume elements (soil, universe) inhabited by the smaller forces *should be* more numerous than the volume elements inhabited by bigger forces. The prediction of a hierarchy of numbers of multi-scale volume elements can be checked, although I think it is obvious. Important is the hierarchy of the tree-like architecture of the universe, which occurs naturally because freedom is a fundamental property that defines the universe.

By the way, one doesn't have to use the language of the infinitesimal (particles and subparticles) to describe the flowing tree of hierarchical forces that fill continuously the outer space. One can look at telescope photographs to see the hierarchy of bodies flowing in space (which, not coincidentally, look similar to Fig. 3.7), and also their arrangements as trees of branching streams, from trunks to filaments. The whole universe is a flowing tree-shaped, interconnected and hierarchical architecture.

References

1. A. Bejan, S. Lorente, A.F. Miguel, A.H. Reis, Constructal theory of distribution of river sizes, Section 13.5, in *Advanced Engineering Thermodynamics*, 3rd edn., ed. by A. Bejan (Wiley, Hoboken, 2006)
2. R.N. Rosa, River basins: geomorphology and dynamics, in *Bejan's Constructal Theory of Shape and Structure* (Évora Geophysics Center, University of Évora, Portugal, 2004)
3. A. Bejan, S. Lorente, A.F. Miguel, A.H. Reis, Constructal theory of distribution of city sizes, Section 13.4, in *Advanced Engineering Thermodynamics*, 3rd ed, ed. by A. Bejan (Wiley, Hoboken, 2006)
4. A. Bejan, S. Lorente, J. Lee, Unifying constructal theory of tree roots, canopies and forests. J. Theor. Biol. **254**(3), 529–540 (2008)
5. A. Bejan, Two hierarchies in science: the free flow of ideas and the academy. Int. J. Des. Nat. Ecodyn. **4**, 386–394 (2009)
6. A. Bejan, Why university rankings do not change: education as a natural hierarchical flow architecture. Int. J. Des. Nat. **2**, 319–327 (2007)
7. A. Bejan, S. Lorente, The physics of spreading ideas. Int. J. Heat Mass Transf. **55**, 802–807 (2012)
8. A. Bejan, P. Haynsworth, The natural design of hierarchy: basketball versus academics. Int. J. Des. Nat. Ecodyn. **7**, 14–25 (2012)
9. A. Bejan, *The Physics of Life: The Evolution of Everything* (St. Martin's Press, New York, 2016)
10. A. Antunes, Encouraging good science on the web. Phys. Today, 41–42 (2009)
11. T.J. Scheff, Academic gangs. Crime Law Soc. Chang. **23**, 157–162 (1995)
12. J.J. Soler, A.P. Moller, M. Soler, Mafia behavior and the evolution of facultative virulence. J. Theor. Biol. **191**, 267–277 (1998)
13. B.R. Clark, The many pathways of academic coordination. High. Educ. **9**, 251–267 (1979)
14. M. Chaput de Saintonage, A. Pavlovic, Cheating. Med. Educ. **38**, 8–9 (2004)
15. News: Italy continues R&D reforms. Nat. Med. **4**(9), 993 (1998)
16. J. Xu, H. Chen, The topology of dark networks. Commun. ACM **51**(10), 58–65 (2008)

17. A. Bejan, Constructal self-organization of research: empire building versus the individual investigator. Int. J. Des. Nat. Ecodyn. **3**, 177–189 (2008)
18. A. Qin, Fraud scandals sap China's dream of becoming a science superpower. The New York Times **13** (2017)
19. A. Bejan, Comment of "Study on the consistency between field synergy principle and entransy dissipation extremum principle". Int. J. Heat Mass Transf. **120**, 1187–1188 (2018)
20. I. Fister Jr., I. Fister, M. Perc, Toward the discovery of citation cartels in citation networks. Front. Phys. **4** (2016), article 49
21. A. Bejan, Letter to the Editor on Temperature-heat diagram analysis method for heat recovery physical adsorption refrigeration cycle—Taking multi stage cycle as an example. Int. J. Refrig. (2018)
22. M. Hoffmann, R. Anderssohn, H.-A. Bahr, H.-J. Weiβ, J. Nellesen, Why hexagonal basalt columns? Phys. Rev. Lett. **115**, 054301 (2015)
23. V.K. Kinra, K.B. Milligan, A second-law analysis of thermoelastic damping. J. Appl. Mech. **61**, 71–76 (1994)
24. L. Levrino, A. Tartaglia, From elasticity theory to cosmology and vice versa. Sci. China: Phys. Mech. Astron. **57**, 597–603 (2014)
25. S.S. Al-Ismaily, A.K. Al-Maktoumi, A.R. Kacimov, S.M. Al-Saqri, H.A. Al-Busaidi, M.H. Al-Haddabi, Morphed block-cracked preferential sedimentation in a reservoir bed: a smart design and evolution in nature. Hydrol. Sci. J. **58**, 1779–1788 (2013)
26. A. Bejan, Y. Ikegami, G.A. Ledezma, Constructal theory of natural crack pattern formation for fastest cooling. Int. J. Heat Mass Transf. **41**, 1945–1954 (1998)
27. A.H. Reis, A.F. Miguel, A. Bejan, Constructal theory of particle agglomeration and design of air-cleaning devices. J. Phys. D Appl. Phys. **39**, 2311–2318 (2006)
28. A. Bejan, R.W. Wagstaff, The physics origin of the hierarchy of bodies in space. J. Appl. Phys. **119**, 094901 (2016)
29. S. Scaringi, T.J. Maccarone, E. Körding, C. Knigge, S. Vaughan, T.R. Marsh, E. Aranzana, V.S. Dhillon, S.C.C. Barros, Accretion-induced variability links young stellar objects, white dwarfs, and black holes. Sci. Adv. **1**(e1500686), 9 (2015)
30. A. Bejan, The constructal-law origin of the wheel, size, and skeleton in animal design. Am. J. Phys. **78**(7), 692–699 (2010)
31. A. Bejan, J.P. Zane, *Design in Nature, How the Constructal Law Governs Evolution in Biology, Physics, Technology, and Social Organization* (Doubleday, New York, 2012)
32. www.fi.edu/lareates/helen-rhoda-quinn
33. H. Georgi, H.R. Quinn, S. Weinberg, Hierarchy of interactions in unified gauge theories. Phys. Rev. Lett. **33**(7), 451–454 (1974)

Chapter 4
Inequality

Hierarchy is often perceived as inequality, which is a divisive impression about us as members of civilized society [1]. The usual argument is that wealth inequality has to increase as wealth increases because contributions to wealth are unequal, meaning that the wealthier contribute more. Reducing inequality calls for redistributive measures, which reduce the incentives for the wealthier to produce wealth. From this, the argument goes and results in an equilibrium between more wealth and more equality. This entire rationale is descriptive, not predictive, and so is the recipe for curtailing inequality. This is why the theoretical (predictive) step advanced in this chapter is timely.

The distribution of wealth should be expected to be nonuniform because it is closely related to the evolutionary movement of all the flowing streams of a live society [2]. What some recognize as inequality, others recognize as hierarchy in an organization alive with movement that has freedom to change. The hierarchical configuring of human movement on the surface of the earth happens naturally. Hierarchy is unavoidable and impossible to efface. Here is why:

Movement that is transferred from individual to individual (by "diffusion") is as omnipresent as tree-channel flow. Flow by diffusion is located between the tree channels on the same area. Diffusion and channels fit hand in glove, like the seepage in the wet banks between river channels. The seepage is the glove; the tree channels are the hands and fingers. This way, everything evolves and flows as one.

The hierarchical movement of people and their belongings and vehicles and communications are physical reality, palpable, and measurable. This physical reality has been known for much longer as human settlements, trade, city living, economics, business and wealth, and government as well. New is that this reality is predictable based on physics.

Physics impresses all of us with its laws, which are few, precise, and universally applicable, valid for any imaginable situation and flow system. Can economics be a more precise and predictive science, like physics [3–8]? Scientists have long recognized the need to discern the physics that underlies the economy [9–13]. The field

A. Bejan, *Freedom and Evolution*,
https://doi.org/10.1007/978-3-030-34009-4_4

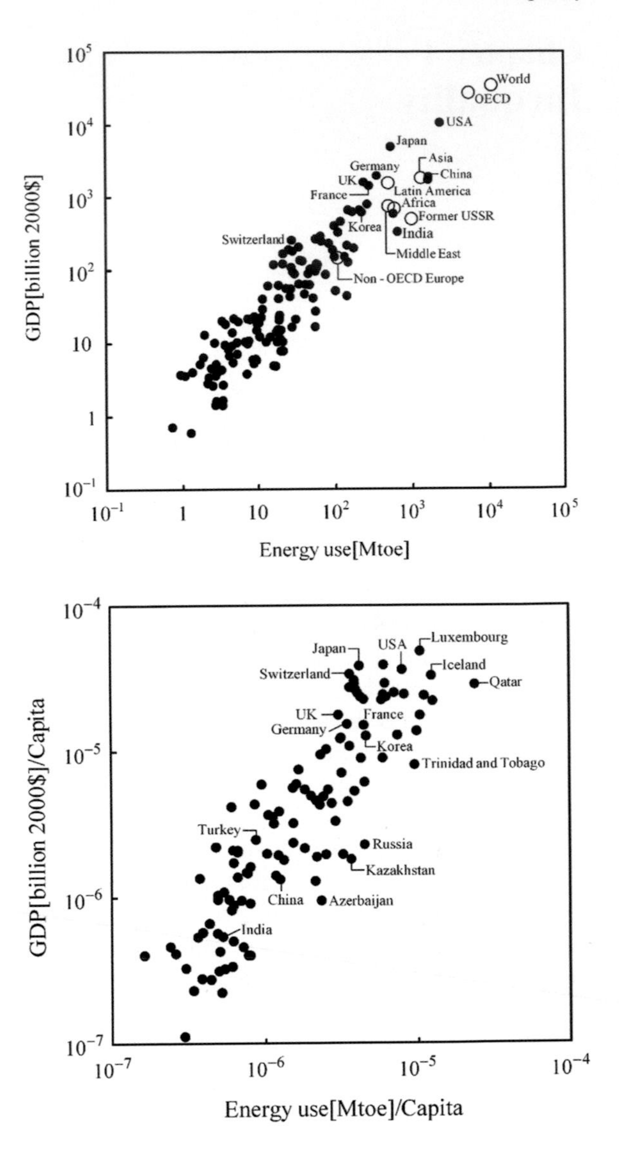

Fig. 4.1 Gross Domestic Product of regions and countries all over the globe versus annual consumption of fuel, aggregate (top) and per capita (bottom). The data are from the International Energy Agency and refer to the year 2006 [33]. Newer reports [26, 34] show that human-related energy expenditure is not uniform over the planet

that grew is known as Econophysics [14]. Not noticed was the need to identify the law of physics that accounts for the tendency toward design evolution everywhere, not just in society.

Recent work based on the constructal law [3–5] has shown that economic activity is closely related to the movement of all the streams of society. The annual domestic economic activity of a country (the Gross Domestic Product, or GDP) is proportional to the physical movement measured as the amount of fuel consumed annually with purpose in that country (Fig. 4.1). Why, because fuel generates power, power drives all the movement, and movement dissipates the power, as shown in Fig. 1.5.

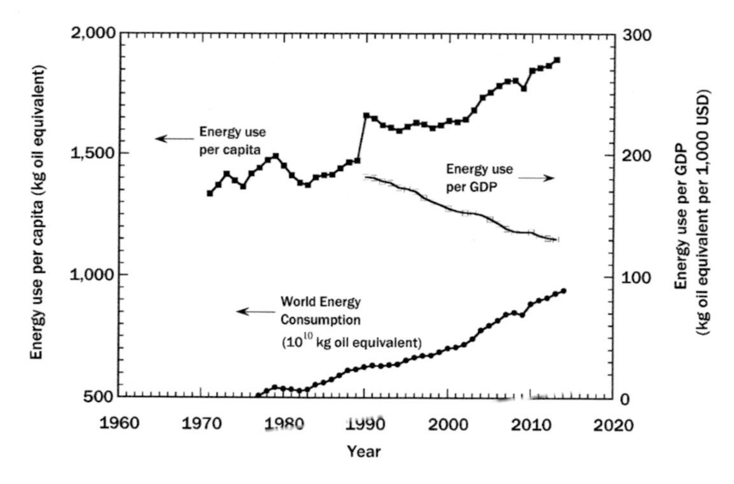

Fig. 4.2 Three ways to look at the evolution of fuel energy use and wealth since 1971. To the left are indicated the annual world energy consumption (the bottom curve) and the energy consumption per capita (the top curve). To the right is shown the energy consumption per GDP (the middle curve). The units are kg oil equivalent per $1,000 GDP, held constant with 2011 international purchase power parity. *Source* IEA Statistics© OECD/IEA 2014—IEA, 2016 and The World Bank Development Indicators, World Bank, 2016 [35]

More economic activity means more fuel consumption, not less. All the dots are racing upward on the diagonal in Fig. 4.1. This evolutionary aspect is documented by the data plotted versus time in Fig. 4.2. Improvements in efficiency (the middle curve) lead to more fuel consumption, not less. The improvements are akin to the removal of obstacles in the flow design, and once removed the obstacles are forgotten. The new design without obstacles flows better, and as a consequence it persists in time. It is adopted.

This consequence serves as answer to an old puzzle in economics known as Jevon's paradox [15]. The puzzle was that the more efficient use of coal in the industrialized world in the 1800s actually increased the consumption of coal and other resources, instead of "saving" them. Figure 4.2 shows that the world energy consumption and the world energy use per capita continue to increase even though the energy use per GDP has decreased since 1990 as the world technology for power generation becomes more efficient.

Before we continue, let's take a step back. The unification of physics and economics advanced here requires unambiguous definitions of the terms that will be used. Words matter.

Movement is the change in the location of material, from one spot to another on the earth's surface. Movement occurs essentially in the horizontal direction. The movement is greater when the mass moved (M) and the distance traveled (L) are greater. The physical measure of movement is the product ML, and so is the expenditure of useful energy (fuel, food), or work spent on causing the movement. Movement is present at all discernible scales, from water in the smallest rivulets to water in the biggest rivers and the trucks on the longest and widest highways.

The physical measure of the movement *ML* is the work spent to cause the movement. The work is the product *FL*, where *F* is the horizontal force overcome by the mover. The product *FL* is proportional to *ML*, because *F* is proportional to *M* in all the media where movement occurs: in water, on land, and in the air [3]. We will return to this observation shortly.

Another physical measure of movement is the amount of fuel (fossil, renewable, food) used for generating the work spent in order to create the movement. This measure is proportional to the first two. The amount of fuel used is proportional to the amount of useful energy (called exergy) contained in the amount of fuel [16]. The spent exergy is proportional to the work produced and consumed during the movement. In the theoretical limit where the energy conversion machine that produces work from fuel operates reversibly, the exergy spent is equal to the work produced and spent during movement. In an actual machine, the generated work is a significant fraction of the exergy content of the fuel consumed. Important is that according to the method of scale analysis [17] the work and the fuel exergy have the same scale, i.e., the same order of magnitude.

In sum, more movement represents more work that is spent, and more work spent represents more fuel that is consumed. Movement is defined as displacement against forces that resist the displacement (Fig. 1.5).

Wealth is a common term that represents the availability of valuable (purposeful) material resources. It is a measure of the assets of value owned by an economic entity. Key is the word "valuable," and in this chapter we zero in on the physical meaning and measure of "value." In the discipline of economics, wealth is described in dollar terms, as on the ordinate in Fig. 4.1. During the past two decades, constructal-law theory showed that the annual wealth of a population or territory is essentially proportional to the useful energy (or work and movement) generated annually by that group, or on that territory. This discovery is summarized in Figs. 4.1 and 4.2, and it is of empirical nature. It is the facts.

Even in the same country, where movement is difficult (e.g., mountain villages) there is less wealth than where movement is easier and better connected (e.g., villages on the plain). Economics and physics are the two sides of the same coin.

This finding is pivotal for science. It means that the economics concept of wealth has a physics basis, which is measurable as work, fuel consumed, or movement effected by fuel, food, and work. This is the unification of economics and physics. The equivalence between wealth and movement is correct in the broad sense, as the scatter in the data in Fig. 4.1 indicates. Outliers exist, and undoubtedly the equivalence is evolutionary because wealth and fuel use are increasing over time, and the dots are migrating upward (one way) along the diagonal.

The evolutionary nature of the alignment shown in Figs. 4.1 and 4.2 includes many effects that are time-dependent. Among these are discoveries of new fuels (fossil and renewable), implementation of new power generation technologies, and national energy policy such as renewables, decarbonization, government, and international trade agreements. See the linkage between individual innovation and broad wealth in the next chapter. Such complexities add even more emphasis to the broad interpretation of Figs. 4.1 and 4.2, which amounts to the translation of the term "wealth" into physical terms and concrete measures (useful energy, work, movement).

Nonuniform distribution is used here descriptively, to account for the physical occurrence of movement (or fuel use, work, wealth) that is not distributed uniformly over a population or a territory. This aspect of nature is widely recognized as hierarchy: few large and many small move, flow, and live together [2, 3, 18, 19]. We see this in the flowing and morphing architectures of river basins, animal populations (the food chain), social organization, streets in the city, highways, the global air traffic, and commerce.

Inequality is an alternative description of the nonuniform hierarchical distribution of movement on earth. This alternative is common when it refers to two equivalent measures of movement, namely, fuel use and wealth. Inequality has a negative connotation implying lack of justice, empathy, and access to wealth. This implication is in total contradiction with the natural origin of hierarchy, which is freedom of movement. The origin of hierarchy lies in the equal access that freedom provides to the whole, to morph its flow architecture, and to liberate its flows.

Even though the origin of hierarchy lies in equal access and freedom, the negative connotation of inequality and injustice cannot be taken lightly. The reason is that justice, or injustice, is in the eye of the beholder. It is a natural feeling that can differ 100% from neighbor to neighbor. Here is an image from a famous Russian story: There are two poor neighbors in a village. One has a goat, the other has none. The envious man kills his neighbor's goat. Both men feel injustice, but it is not the same.

The talk about justice is common but loaded with the personal preconceptions of every individual who speaks the word. It is divisive. The word justice (from the Latin *justus*, the quality of being right, correct, as in to adjust this to fit that) does not mean what the politicians mean. I would not elevate the word justice to describe a sense that supersedes that of individuals. The name for what supersedes all individuals is social organization, which happens naturally, not because the individuals trade in "justice" among themselves until they all arrive at (and agree with) a design of life flow on the area. Most people are never consulted about anything. They vote with their wallet and with their feet. They are individualistic: they join, because it is good for them, individually. Good for them is access and ease of flowing in all the ways that matter in life.

Let's consider the movement of anything on earth. In the simplest terms, movement on the earth's surface is measured as the amount of fuel spent in order to drive the movement. The fuel spent is proportional to FL/η, where F is the horizontal force of resistance overcome by the mover, L is the distance traveled, and η is the energy conversion efficiency of the engine that drives the mover. According to Figs. 1.4 and 1.5, the efficiency of the engine is the ratio $\eta =$ (work output)/(heat input), which is proportional to the ratio (work output)/(fuel used).

According to the physics of locomotion on earth in all media (water, land, air) [3, 20], the horizontal force F is equal (in an order of magnitude sense) to rMg, where g is the gravitational acceleration, M is the mass moved horizontally, and r is

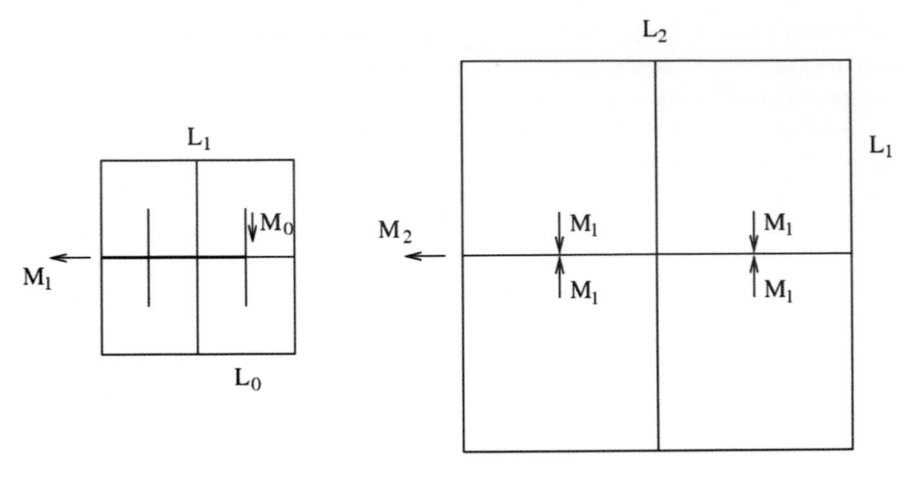

Fig. 4.3 Movement on an area as a sequence of progressively larger constructs formed by quadrupling

a dimensionless factor that accounts for the medium in which the movement occurs ($r \sim 1$ for swimming, $r \sim 0.1$ for flying, and $0.1 < r < 1$ for running).

At the same time, the physics phenomenon of economies of scale (Chap. 2) dictates that larger engines must be more efficient. This means that the efficiency η increases slowly with the engine size M, for example, in proportion with M^{α}, where the exponent α is a positive number smaller than 1, such that the curve $\eta(M)$ rises at a decreasing rate. In a mature technology, the improvements in efficiency by changes in the design are marginal (cf., Chap. 10), η tends toward a plateau, and consequently α is less than 1. One example ($\alpha \sim 1/4$) is the convergent evolution of the engines for helicopters, Fig. 2.4.

In sum, the amount of fuel spent during the movement of the mass M to the distance L on earth is proportional to the product $M^{1-\alpha}L$, where the exponent $(1 - \alpha)$ is a number smaller than 1 because α is smaller than 1.

Consider next a large territory (for example, a country) swept by movers of many sizes ($M_0 < M_1 < M_2 < \ldots$) and unequal numbers. The movement on the territory is a modular (constructal) flow architecture (cf., Fig. 3.2), where each larger construct is made of a number (n) of smaller constructs. This construction is illustrated here in Fig. 4.3 for $n = 4$, which is an architecture based on quadrupling as in the theoretical construction of river basins of all sizes [21]. Other rules of construction ($n = 2, 3, 6$) lead to essentially the same conclusions, as we will see in Figs. 4.4 and 4.5.

The smallest construct is a square of side L_1. There are four small movers (M_0) that feed their freight to one large mover (M_1), which exits the $L_1 \times L_1$ area. There are N_1 such areas that make up the whole territory. The M_1 mover travels a distance comparable with L_1. Each M_0 mover travels a distance comparable with $L_1/2$. At the next larger scale, four movers of size M_1 feed their freight to one mover (M_2) that exits the territory of size $L_2 \times L_2$, in which $L_2 = 2L_1$. There are N_2 areas of size L_2

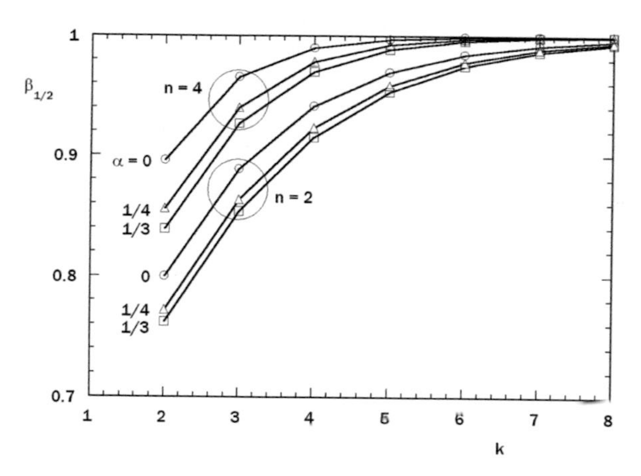

Fig. 4.4 The movement (wealth) is distributed nonuniformly among the movers. The total wealth is divided equally between two groups, the wealthier and the less wealthy. The fraction that the less wealthy represent in the total population is reported as $\beta_{1/2}$ versus n and k. The less wealthy represent a much larger fraction of the total population than the wealthier. The lower graph shows the effect of the economies of scale exponent α. The unequal distribution of wealth becomes accentuated as the complexity (n, k) increases, and the movement covers the territory more completely

$\times\ L_2$ that fit in the country. The rules of this construction toward larger constructs are evident: $L_{i+1} = 2L_i$, $N_{i+1} = N_i/n$, and $M_{i+1} = nM_i$.

The total fuel spent (i.e., the wealth W_k) is proportional to the sum $W_k = N_1\left[M_1^{1-\alpha}L_1 + nM_0^{1-\alpha}L_0\right] + \cdots + N_k\left[M_k^{1-\alpha}L_k + nM_{k-1}^{1-\alpha}L_{k-1}\right]$ where k is the number of levels of construction. Only two levels of construction are shown in Fig. 4.3, namely, $k = 1$ and $k = 2$. In view of the rules of construction, the total movement on the territory becomes $W_k = f\ N_1M_0^{1-\alpha}L_0(\gamma^k - 1)/(\gamma - 1)$, where $\gamma = 2n^{-\alpha}$ and $f = \frac{1}{2}n^{2-\alpha} + n$. The details of this analysis are available in Ref. [2].

The W_k sum represents the total movement over the constructs interconnected at all the scales, that is, over the k levels of construction. This total is also proportional to the total wealth, as we saw in Fig. 4.1. On the same territory, the population of movers (counted in the direction from the many small to the few large) is $P_k = n^{k-1} + n^{k-2} + \cdots + n + 1 = (n^k - 1)/(n - 1)$.

Next, we ask whether the total movement W_k is distributed evenly over the population P_k. The area constructs are numbered from the smallest to the largest: 1, 2, …, k. Imagine the intermediate size represented by the size of the construct level j, which accounts for all the movers of size M_j and smaller. Their combined movement $W_j = f\ N_1M_0^{1-\alpha}L_0(\gamma^j - 1)/(\gamma - 1)$ represents a fraction (ε) of the total, $\varepsilon = W_j/W_k = (\gamma^j - 1)/(\gamma^k - 1)$, $(0 < \varepsilon < 1)$. The fraction (β) of the total population that accounts for the movement fraction ε is $\beta = P_j/P_k = n^{k-j}(n^j - 1)/(n^k - 1)$, $(0 < \beta < 1)$.

The construct of size j divides the total population into two groups, P_j and $(P_k - P_j)$. If the total movement is divided equally between the two groups, that is, if $W_j = \frac{1}{2}W_k$ or $\varepsilon = 1/2$, then we can use $\varepsilon = W_j/W_k = 1/2$ to determine the construct

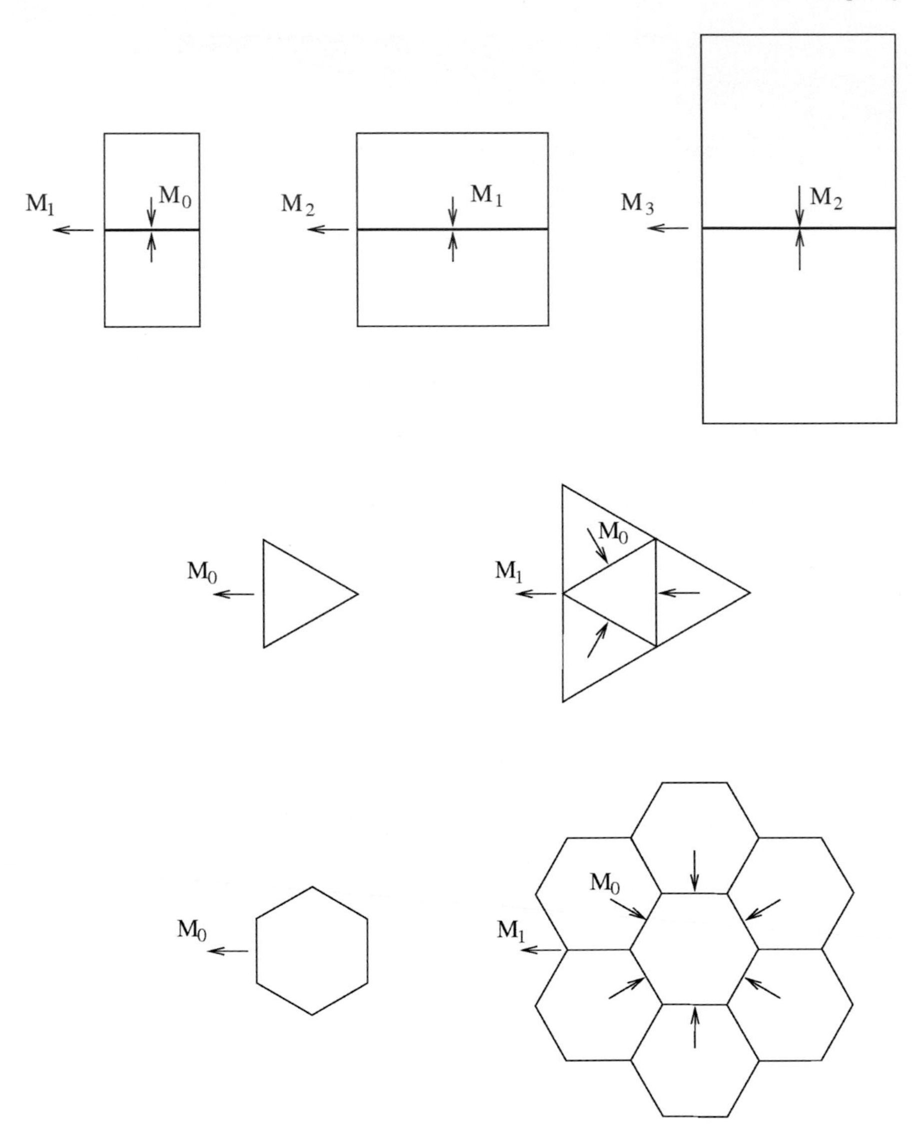

Fig. 4.5 Movement on areas constructed by doubling, tripling, and multiplication by six

of size j that divides the movement (wealth) between the two groups equally, smaller movers (constructs 1, 2, ..., j) versus larger movers ($j + 1$, ..., k).

The population fraction that belongs to individuals with less movement is obtained by substituting the calculated j into $\beta = P_j/P_k$. The resulting fraction is labeled $\beta_{1/2}$ on the ordinate of Fig. 4.4. The subscript 1/2 indicates the equipartition of movement between the two population groups. Figure 4.4 was drawn by assuming $n = 4$ and $n = 2$, and by varying the exponent α in the vicinity of 1/4. The individuals with less movement are the less wealthy who are thought to represent 80% of the total population, as observed by Pareto [22] and several generations of economists since. This means that the model and analysis that led to Fig. 4.4 indicates that for a fixed $\beta_{1/2}$ value such as 0.8 there must be a relation between n and k, for example, $n = 2$ if $k = 2$ or 3. The unequal distribution of movement (wealth) becomes more accentuated as the economy becomes more developed, i.e., as its flow architecture becomes more complex for the purpose of covering progressively smaller interstices of the overall territory, which is fixed.

Many people are unaware of the hierarchical area movement of the society. This is why the movement of water in a river basin serves as reference, because it is familiar to all of us. Since the 1930s, the hierarchy of rivers and tributaries in a river basin is being documented in geophysics and summarized as empirical rules (improperly called "laws") named after Horton [23], Melton [24], and Hack [25]. The best known rule is Horton's law of stream numbers, which is based on counting and measuring the channels of many river basins. The rule that emerged from measurements is that the number of tributaries to a larger river falls in the range between 3 and 5. Models of river basins assign a constant value to this number, even though there is no real-world river basin with a universal (an integer) number of tributaries.

We develop a view of the natural phenomenon of hierarchical movement by imagining a river basin that covers a fixed area with a dichotomous design ($n = 2$, Fig. 4.5) where every channel has two identical tributaries. The movement (the water flow rate) in one channel equals the movement in the tributaries. Next, we reason as follows:

(i) If the area is covered by only three channels, then the architecture is Y-shaped, with one level of construction ($k = 1$). The movement in the big channel equals the movement in the remaining two channels, therefore $\beta_{1/2} = 2/3$.
(ii) If the basin has seven channels, the architecture is more complex ($k = 2$), the flow in the big channel equals the flow in the remaining six channels, and $\beta_{1/2} = 6/7$.
(iii) If the architecture is even more complex, with 15 channels, then $k = 3$, and $\beta_{1/2} = 14/15$.

In the sequence (i)–(iii), we discover that it takes only a modest degree of complexity ($k = 2$ or 3) for the measure of inequality ($\beta_{1/2}$) to reach levels even higher than the 80% observed by Pareto. This conclusion is the same for constructions based on doubling (Fig. 4.5 top) and quadrupling (Fig. 4.3). The lower part of Fig. 4.5 suggests two alternative constructions, based on tripling and on compounding areas by six at a time.

The scales of the constructions made in the models of Figs. 4.3 and 4.5 have lower and upper bounds that are finite. Finiteness is an intrinsic feature of the evolutionary design phenomenon captured by the constructal law: the use of "finite size" in the constructal-law statement (page 5) rules out the infinitesimal and the infinite. In a river basin, the lower bound is the area of the two hill slopes that feed the first rivulet, and the upper bound is the down-slope territory between the mountain range and the coastline. In urban design, the lower and upper bounds are the area occupied by one

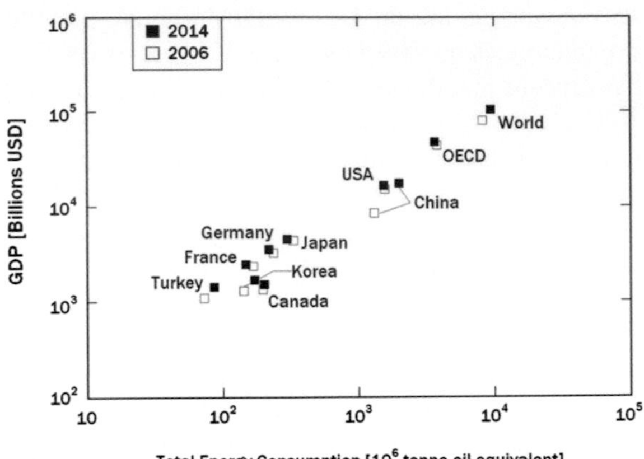

Fig. 4.6 The upward migration of the data of Fig. 4.1, from 2006 to 2014 [26, 33–35]

house (covered by walking) and the area of the city (covered by riding on one or more vehicles).

The chief conclusion is that "inequality" is part of the universal phenomenon of evolution in nature. On this physics foundation, it pays to re-examine the observations presented in Figs. 4.1 and 4.2, which served as starting point for the unification of economics with physics:

First, in Fig. 4.1, every point—every population group—is racing upward. This is now validated by Fig. 4.6, which focuses on the wealthy end of the data of Fig. 4.1 and shows how the data have migrated upward from 2006 to 2014. The migration is consistently upward, and it is of two kinds. Developing countries such as China and Turkey have migrated upward along the diagonal. Developed countries have migrated upward and to the left. This second group has become not only wealthier but also more efficient, by creating more movement per unit of fuel spent.

The new aspect revealed by Fig. 4.6 is that the evolutionary design is such that the tendency toward greater movement goes hand in hand with the tendency toward more efficient technologies that enable the movement. This elucidates Jevon's paradox. Both tendencies are manifestations of the constructal law because this combination is for easier movement and greater access.

Second, data from the World Bank [26] offer additional support for the natural occurrence of unequal distribution of wealth. On the ordinate of Fig. 4.7, we see the share (ε) of the annual income of the population fraction (β). The comparison between Figs. 4.7 and 4.4 is not exact because income (Fig. 4.7) is not exactly the same as wealth (Fig. 4.1). Nevertheless, the comparison is qualitatively valid, and it shows that income is distributed unequally everywhere. The distribution curve is convex. Consequently, the present theory also predicts the curve of distribution of income inequality, which has been recognized empirically for a century.

The diagonal in Fig. 4.7 corresponds to a population in which wealth is distributed equally, namely, $\varepsilon = \beta$, or, for example, 40% of the population owns 40% of the wealth. Such distributions were the design in prehistory (long before antiquity), when the area elements inhabited by hunters and gatherers were self-standing, sparse and

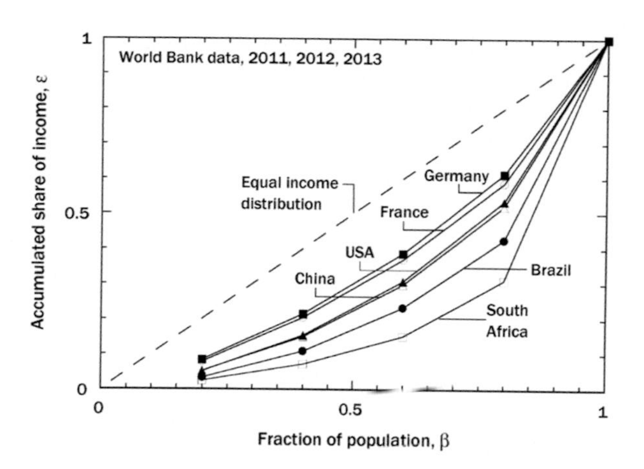

Fig. 4.7 The accumulated share of income ε held by the fraction of the population (β). Data adapted from Ref. [26]

disconnected settlements, without social organization, work animals, roads, and trade linking them. Equality was everywhere, all right, and so was poverty.

There is more to the prehistoric design represented by the diagonal in Fig. 4.7. With the lack of physical movement came not only the poverty but also the lack of impact on the environment. With poverty comes unawareness of "environment" (recall that movement means having impact, i.e., getting the environment out of the way). Prehistoric humans were not "protecting" the environment—that was not their concern. The opposite was happening. Early humans, like all animals, engaged in "niche construction." They were innately oriented toward sculpting their "niche," and changing and harnessing the environment to make human life easier and longer lasting, to survive.

In advanced societies today, some are championing both features, the equality and the environment, with nostalgic references to the "equality" and "respect" for the environment that seem to characterize the poorest living groups on the globe. Marxism has been championing equality and spreading communal life for 150 years. The physical reality is that both features, lack of hierarchy and lack of environmental impact, belong to populations with little movement, grave poverty, hunger, and short lives.

Figure 4.7 further shows that the more advanced countries have relatively straighter (less convex) curves, i.e., their income is distributed somewhat more equally. This new aspect is an invitation to expanding the "economics as physics" theory into the domain of social organization (Chap. 5). All distributions are happening naturally, yet their degree of "inequality" can be tweeked with design features (e.g., philanthropy, taxation) that so far are not present in the theory advanced in this chapter.

To develop a feel for the physical effect of taxation, consider the river basin as a metaphor. The river equivalent is to put channels in places where channels did not happen by themselves, and to spend power (fuel) to force the flow (e.g., freight) through those channels. This artificial design comes on top of the natural design, and it happens only in the advanced (wealthy) countries, where saved power (known as money saved in banks) is available to be spent. The artificial are the human-made canals with locks, which link the natural rivers and wet the previously dry areas between rivers. For the new traffic to happen, power must be spent to build the canal, to maintain it, and to operate the system of locks.

There are many more examples of how to construct channels where they do not exist, and they are known by many names. Most inspiring to me are the voluntary acts. World leading universities were paid for by private individuals such as Carnegie, Duke, Vanderbilt, and Rockefeller, which are now permanent factories (assembly lines) of artificial channels for young people who otherwise would be stuck, not knowing, not moving, not advanced, and not wealthy. I think of these great men every time I get up and go teach in Africa and other faraway places, where I give students a chance to join my world.

Channels that illustrate in physical terms the beneficial effect of movement are the Canal du Midi in southwest France, the canals in Britain and northeast United States, and the Suez and Panama canals as well. People tend to think of canals only in terms of facilitating transportation. That makes sense, yet it obscures the increased movement and wealth that are associated with the new design. The people who live where the canals are being built are instantly (stepwise) uplifted to prosperity, compared to who they were when their land was dry and not connected to anything that flows big. The whole globe gets richer because of the movement liberated in one spot (cf., innovation, Chap. 5).

The push for STEM education is a good example of building channels where they do not happen by themselves. Much more numerous, effective, and long lasting are the flow channels built quietly by private individuals. Additional channels are good, the more numerous the better. With them comes a more resilient and richer texture of the fabric of live society.

Another starting point for further work along this theoretical line is the observation that the population is increasing at all levels, on finite areas of many sizes. This introduces a time-dependent aspect to the inequality in wealth and movement distribution. Population growth may also contribute to the physics that underpins Jevon's paradox, as the increase in fuel consumption is related to the increase in fuel demand caused by population growth. The reverse is also true because with more fuel consumption (with more power) the global flow system of human life increases and spreads in accord with its S-curve history and future [27]. The S-curve phenomenon means that every point-area flow spreads over its territory in three successive phases, slow–fast–slow. This is why world population, fuel consumption, and movement (such as miles driven in the U.S. annually) are reaching the plateau. They cannot increase indefinitely.

Fuel is being consumed with purpose, wisely, not aimlessly. How do we know this? Hidden under the global curves shown in Figs. 4.4 and 4.7 are the actual flow paths of human movement. The paths are constantly morphing to provide easier access to what flows. The tendency to evolve and improve the paths traveled by people and goods on the globe (to make them more economical) is the object of geographical economics [28–30]. This tendency is also known as the law of parsimony. It governs

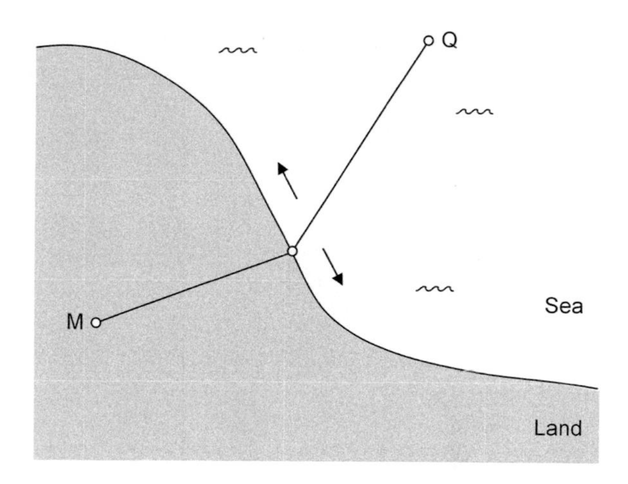

Fig. 4.8 The refracted path traveled by goods from sea (Q) to land (M), and the natural emergence of the harbor, as the economical point of loading and unloading on the shore. Before the emergence of "the harbor," all the locations on the shore were eligible for consideration by those transporting goods from sea to shore and from land to shore

human decisions of where to place routes on earth, and where to construct points for loading and unloading.

In the example of Fig. 4.8, goods are transported by land and by sea between two points that are fixed, M and Q. The point of loading and unloading ships is free to slide on the coast, and it settles in a place that makes the movement between M and Q more economical. The line connecting M with Q is one broken line—one refracted ray—because land transport is more expensive per unit than sea transport.

The emergence of the harbor on the coast is natural, to facilitate the flow from land to sea and back to land. The harbor grows while the flow, commerce, economy, and wealth grow. The natural origin of the harbor contradicts the saying that a city grew because it happened to be at the intersection of trade routes. No, the trade was from one point M to one point Q when there were no routes. Natural was the birth of a single refracted route, not the intersection of existing routes. The location of the refraction point is the manifestation of the universal tendency of evolutionary flow design.

Double refraction also happens. Figure 4.9 shows the tendency toward reduced transportation costs between two points, Hawaii and New Orleans. This tendency is the cause of bends in the route when the goods must be transported not only by ship but also overland [31]. The broken line of the point-to-point route is the law of refraction in economics and a manifestation of the constructal law.

Bundles of such routes make up the tree-shaped paths that connect points (distributors and collectors) with large and small territories (very large numbers of individual consumers, or producers). The trees of our economies, local and global, are generated by the same global tendency that produced Fig. 4.9 and all the other paths traveled by humans on the world map.

Refraction of paths for easier flowing is everywhere you look, especially at human scales, which are much smaller than what can be plotted on the world map. In the construction of a pile of stones such as an ancient pyramid (Fig. 4.10), each stone follows a refracted path. On the horizontal is the easier movement, by sliding. On the

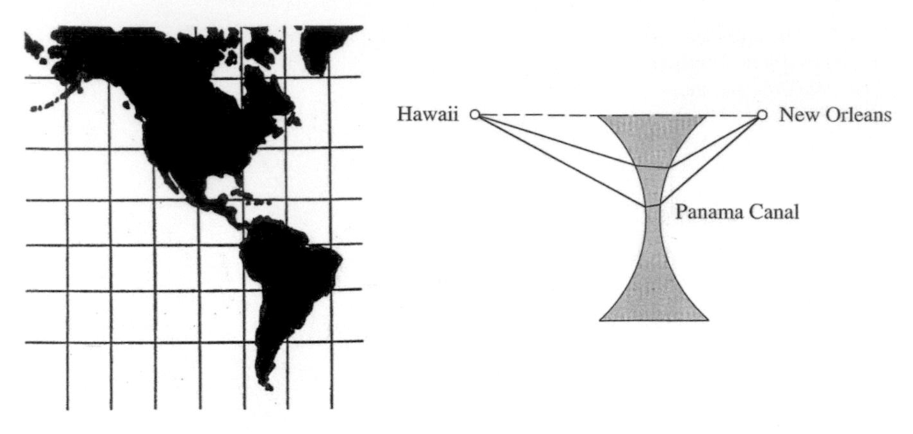

Fig. 4.9 The refracted path for more economical transportation between two points on the globe

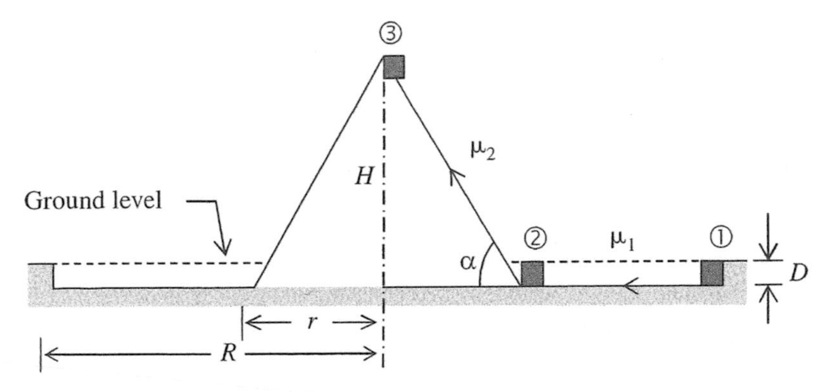

Fig. 4.10 The constant-angle growth of a pyramid. Every stone follows a path that is a refracted ray, such that the global movement of stones requires minimal effort

incline is the more laborious movement, by hopping each stone on the steps created by the previous stones. The angle of refraction (α) is free to vary, yet, that angle is the same in all the pyramids, from Egypt to Central America [32]. It is this way because it is a manifestation of the universal tendency of all flow designs that evolve freely toward providing easier and greater access.

At bottom, evolution happens on the same physics basis in economics, geophysics, animate, and inanimate systems. Evolution is one name for the changes that occur in a flow architecture in a discernible time direction. The design change that facilitates global flow is adopted and persists. It survives to live another day and get better. The evolution phenomenon is evident in economics.

The division of labor and the invention of money were dramatic changes for facilitating the flow of traded goods, big changes relative to trading in nature. The adoption of a single currency in the USA, and more recently in the eurozone, conveys the same message. The spreading of literacy, English language, Roman script, free trade agreements, and the Internet lead to increased movement.

Evolution unites physics with economics and extends the reach of the physics law of evolution over the phenomena of social organization and movement on the globe. The fact that inequality is natural (i.e., physics) has broad implications for the future of western society and capitalism, and the avoidance of class hatred and war.

Less inequality might lead to less envy, less hatred, less violence, and more peace, but no inequality is impossible. What's left is a continuation of the natural order of things that happen by themselves, and which have empowered western society to this day: work ethic, philanthropy, merit system, questioning authority, rule of law, change, hierarchy, and above all freedom.

References

1. W. Scheidel, *The Great Leveler* (Princeton University Press, Princeton, NJ, 2017)
2. A. Bejan, M.R. Errera, Wealth inequality: the physics basis. J. Appl. Phys. **121**, 124903 (2017)
3. A. Bejan, *The Physics of Life: The Evolution of Everything* (St. Martin's Press, New York, 2016)
4. G. Lorenzini, C. Biserni, The Constructal law: from design in nature to social dynamics and wealth as physics. Phys. Life Rev. **8**, 259–260 (2011)
5. T. Basak, The law of life: the bridge between physics and biology. Phys. Life Rev. **8**, 249–252 (2011)
6. W.M. Saslow, An economic analogy to thermodynamics. Am. J. Phys. **67**(12), 1239–1247 (1999)
7. P. Kalason, *Épistémologie Constructale du Lien Cultuel* (L'Harmattan, Paris, 2007)
8. H. Temple, *Théorie générale de la nation, L'architecture du monde* (L'Harmattan, Paris, 2014)
9. M. Smerlak, Thermodynamics of inequalities: from precariousness to economic stratification. Cornell University Library. arXiv:1406.6441 [physics.soc-ph] (2014)
10. M. Karpiarz, P. Fronczak, A. Fronczak, International trade network: fractal properties and globalization puzzle. Cornell University Library. arXiv:1409.5963 [physics.soc-ph] (2014)
11. G. Weisbuch, S. Battiston, From production networks to geographical economics. J. Econ. Behav. Organ. **64** (2007). https://doi.org/10.1016/j.jebo.2006.06.018
12. Y. Chen, Maximum profit configurations of commercial engines. Entropy **13**, 1137–1151 (2011)
13. W.C. Frederick, *Natural corporate management: from the big bang to Wall Street* (Greenleaf Publishing, Sheffield, UK, 2012)
14. P. Mirowski, *More Heat than Light* (Cambridge University Press, Cambridge UK, 1989)
15. J.B. Alcott, Jevon's paradox. Ecol. Econ. **54**, 9–21 (2005)
16. A. Bejan, *Advanced Engineering Thermodynamics*, 4th edn. (Wiley, Hoboken, 2016)
17. A. Bejan, *Convection Heat Transfer*, 4th edn. (Wiley, Hoboken, 2013)
18. A.F. Miguel, The physics principle of the generation of flow configuration. Phys. Life Rev. **8**, 243–244 (2011)
19. A.H. Reis, Design in nature, and the laws of physics. Phys. Life Rev. **8**, 255–256 (2011)
20. A. Bejan, J.H. Marden, Unifying constructal theory for scale effects in running, swimming and flying. J. Exp. Biol. **209**, 238–248 (2006)
21. A. Bejan, S. Lorente, A.F. Miguel, A.H. Reis, Constructal theory of distribution of river sizes, Section 13.5, in *Advanced Engineering Thermodynamics,* ed. by A. Bejan, 3rd edn. (Wiley, Hoboken, 2006)
22. V. Pareto, Cours d'Économie Politique, vol. II ('The Law of Income Distribution') (1897), in *The Economics of Vilfredo Pareto*, trans. R. Cirillo (Frank Cass and Co., 1979), pp. 80–87

23. R.E. Horton, Drainage basin characteristics. EOS Trans. AGU **13**, 350–361 (1932)
24. M.A. Melton, Correlation structure of morphometric properties of drainage systems and their controlling agents. J. Geol. **66**, 35–56 (1958)
25. J.T. Hack, Studies of longitudinal profiles in Virginia and Maryland, USGS Professional Papers 294-B, Washington DC (1957), pp. 46–97
26. The World Bank, World DataBank. http://databank.worldbank.org/data/home.aspx. Visited on 29 December 2016
27. A. Bejan, S. Lorente, The constructal law origin of the logistics S curve. J. Appl. Phys. **110**, 024901 (2011)
28. A. Lösch, *The Economics of Location* (Yale University Press, New Haven, CT, 1954)
29. P. Haggett, *Locational Analysis in Human Geography* (Edward Arnold, London, 1965)
30. P. Haggett, R.J. Chorley, *Network Analysis in Geography* (St. Martin's Press, New York, 1969)
31. A. Bejan, S. Lorente, Thermodynamic optimization of flow geometry in mechanical and civil engineering. J. Non-Equilib. Thermodyn. **26**, 305–3554 (2001)
32. A. Bejan, S. Périn, Constructal theory of Egyptian pyramids and flow fossils in general, Section 13.6, in *Advanced Engineering Thermodynamics,* ed. by A. Bejan, 3rd edn. (Wiley, Hoboken, 2006)
33. International Energy Agency, Key world energy statistics (2006)
34. International Energy Agency. http://www.iea.org/statistics/. Visited on 29 December 2016
35. IEA Statistics© OECD/IEA 2014. http://www.iea.org/publications/publication/energy-statistics-manual.html

Chapter 5
Social Organization and Innovation

Why does social organization happen by itself? Why does it evolve as the movement of the members of the society increases? Why does the organization become more hierarchical, with greater nonuniformity in the distribution of movement over the inhabited territory when the size of the society increases?

The preceding chapters empowered us to address these questions from physics. Key is the idea that the phenomenon of social organization is rooted in the universal physics phenomenon of economies of scale, in the presence of freedom to change. It takes less power to move a unit mass in bulk, along with many units, than to move one unit alone. The freedom to change and to choose between competing changes is essential.

The physics aspect is crucial because many people believe that human life is too complicated to be predictable. That is true if one attempts to predict the single acts of a single individual. Brownian motion is also too complicated if one attempts to predict the behavior of one molecule. The whole society (the macro) has its own behavior, and that is predictable—the social, the Brownian, the river basin, and the spreading plague. The constructal law of physics covers them all.

Freedom enables the mover to opt for changes that enhance access and facilitate flow. Life is movement, and to live (to persist in time) every individual has the urge to move more easily by morphing its path and rhythm through its environment. Joining and organizing happen as a consequence. In transportation, the fuel spent per unit of freight is smaller on a large vehicle than on a small one. In animal locomotion, the food eaten per unit of animal mass moved is smaller for the elephant than for the antelope. This physics phenomenon is measurable and put on display in many studies of the evolution of animal design and power plants, which show that the efficiency increases with the size of the flow system (cf., Chap. 2).

In this chapter, we illustrate the natural phenomenon of joining (organization) with two highly dissimilar models [1]. One model is inanimate: river basins generated by several rules of construction (Fig. 9.2). The other model is also about water flow, but

A. Bejan, *Freedom and Evolution*,
https://doi.org/10.1007/978-3-030-34009-4_5

it is animate: the production and distribution of heated water over human settlements that grow on an area.

The question is whether the phenomenon of social organization can be placed on a physics foundation. This question is stimulating a growing domain in physics, with visible progress in crowd dynamics, economics, urban design, and cultural evolution. The physics literature reviewed in Ref. [1] and cited in the preceding chapters shows that a physics basis exists for phenomena and "intangibles" previously associated with the social realm, for example, economies of scale, diminishing returns, wealth distribution, hierarchy, the living space, and evolution as physics.

To get to the physics basis of social organization, it pays to review the core lessons of the preceding chapters: Nothing moves unless it is forced to move against an environment that resists. The forcing comes from power; the power comes from fuel and food consumed in "engines," which are natural and human-made, and the movement evolves with freedom toward architectures that offer easier and greater access to what flows.

Furthermore, on a finite area or in a finite volume, the evolved flow architectures are tree-shaped, hierarchical, with few large and many small channels flowing together to bathe the whole flow system (river basins, lungs, snowflakes, city, and air traffic). In Chap. 4, we saw that the annual consumption of fuel in society is proportional to the annual measure of wealth. Because the movement on a territory is hierarchical, and because movement is the physical consequence of the consumed fuel, the consumption of fuel is also spread nonuniformly and hierarchically on the globe. Finally, because fuel consumed is proportional to wealth, the unavoidable conclusion from physics is that wealth must also be distributed hierarchically ("unequally") in the live, flowing and freely morphing world.

The predictions that result from this physics theory are in accord with observations throughout the recorded history of civilization. Hierarchy and wealth inequality "happen," while any effort to eliminate inequality completely is short lived, marginally successful at best, violent, and murderous at worst. The river basin has its physics basis, and because of the physics the architecture always emerges as hierarchical. Likewise, a flowing and thriving society is eminently hierarchical.

Caution: what I just wrote should not be misrepresented to mean that there is no point in trying to curtail inequality. Any society tries that naturally, because it is good for the peace, the life and the persistence of the whole. The message from physics is that to *eliminate* inequality is impossible. Every individual and society is better off knowing the difference between the possible and the impossible: the difference is explored further in Chap. 10. Everyone is better off knowing what to try and what not to try. This difference is the key to envisioning, designing, and implementing one's future.

The clue to unearthing the physics basis of social organization is shown in Fig. 4.7, in which income inequality is indicated by the gap between the diagonal and each of the six curves. Intriguing is the impression that the society of South Africa is not making a strong effort to reduce inequality. The impression is wrong. Here is how to check the validity of this impression. Is every social group acting toward a more uniform distribution of wealth? If the answer is yes, then this is the

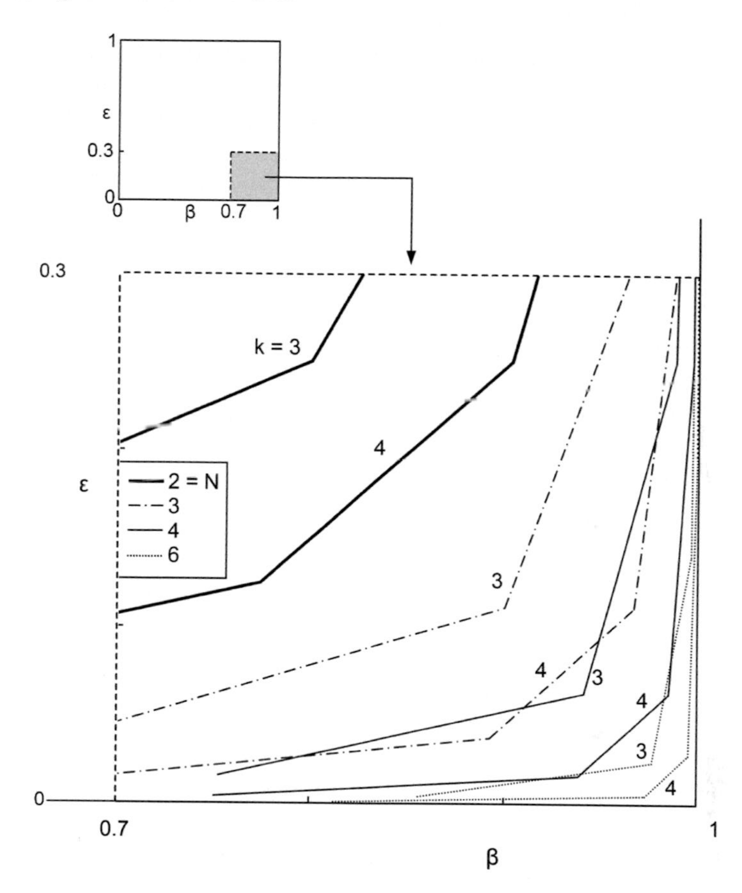

Fig. 5.1 The hierarchical distribution of flow rate (ε) in the population of channels (β) in the river basin of Fig. 4.3. The figure shows that most of the population (large β), accounts for a small fraction of the total flow (small ε)

physics basis of social organization with controlled inequality, which happens to be the defining feature of human society today.

The way to approach the question is to start from a reference design, where social organization is completely absent, as in river basins and other "vascular" architectures not made and maintained by humans, either bio or non-bio.

River basins are freely evolving flow architectures. As we saw already, river basins obey rules of construction that hold for all basin sizes, for example, four tributaries (M_1) to one mother channel ($n = 4$, on an average). This construction is sketched in Fig. 4.3, where each of the four tributaries comes from an elemental square area.

In the river basin, the movement is the water flow rate over the territory. Figure 5.1 shows the fraction of the total movement (ε) versus the fraction of the total population of river channels.

Compared with Fig. 4.7, the distribution of movement in river basins is severely nonuniform. In Fig. 5.1, the ε–β curves are crowded in the lower right corner. The hierarchy of this inanimate flow architecture is extreme. Perhaps in a thriving society things will be different.

Why does organization happen in a society in the first place? Throughout the history of social organization, the evolution away from the uniform distribution of flow over the population is a consequence of the trade-off between (a) the generation of the original flow, which is less wasteful when the original stream is larger (because of economies of scale), and (b) the route of transport from source to user, which is more wasteful when the length of the route is greater. Both losses, (a) and (b), are losses in the physics sense: they are measurable in terms of destruction of useful energy (exergy, joules), or waste of fuel and food (kilograms). Both losses increase with the size of the original stream that reaches every member of the population; however, (a) increases more slowly than (b).

The evolution from the "one size fits all" to hierarchy in the human realm can be predicted by considering the production and spreading of one material that enhances human movement (life), for example, a material that was made possible by the control and adoption of fire. Consider the use of hot water, which is essential for better food, more food (food preservation), shelter (warmth), hygiene, health, and every other essential feature of civilized living.

In the early days, water was heated in one vessel for the inhabitants of one settlement, as in the lower left design of Fig. 5.2. The round area element is the territory of one settlement. The point in the center is the settlement and its source of hot water. The amount of hot water used by one settlement per unit time is m_1. The use of hot water, like the burning of wood, was distributed equally over the population.

Over time, users formed a cluster around one central heater that delivered hot water to everyone, radially. Today, the modern distribution consists of increasingly more complex hierarchies with intermediate nodes of distribution between the one large center of production and the many small and equal peripheral users. Why should this have happened?

Complex hierarchies are understood more simply as tree-shaped flows, Fig. 5.3. Their complexity increases over time as the society becomes more advanced, the individual use of hot water (m_1, or $\tilde{m}$) increases, and the number of individuals who join the group increases. On the ordinates of Figs. 5.2 and 5.3, we see the loss of heating (or the wasteful consumption of fuel) during the production of the unit of hot water plotted on the abscissa. These losses occur during heating the water, and along the distribution lines. On the abscissa, the value of $\tilde{m}$ (dimensionless) is proportional to the amount of water needed by one individual per unit time, m_1. The analytical details behind these figures are available in Ref. [1].

As civilization develops, the amount of hot water used by one individual increases. The time axis points from left to right in Figs. 5.2 and 5.3. Over time, it becomes possible to produce and distribute hot water in central heaters that serve clusters of users. In the pursuit of less fuel waste ($\tilde{q}$) per unit of hot water delivered to the population, the arrangement of users and heaters must evolve stepwise, from uniformity ($N = 1$) to larger clusters ($N = 3, 6, \ldots$) with longer distribution lines, as one can

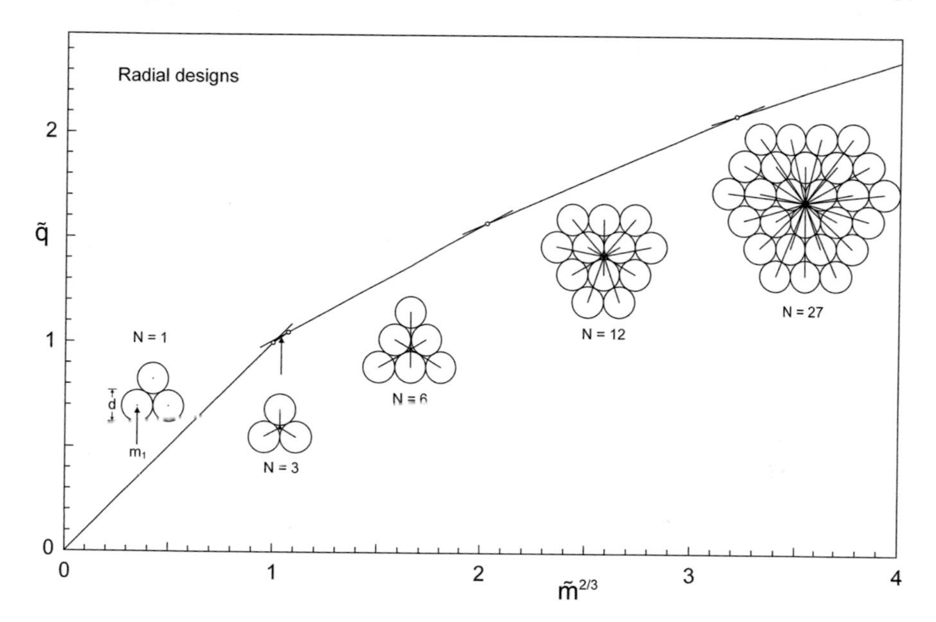

Fig. 5.2 The loss of heat per user versus time ($\tilde{q}$, on the ordinate) versus the amount of water consumed by one individual ($\tilde{m}$, on the abscissa). From left to right (i.e., in time), the organization evolves from individual production and use of hot water ($N = 1$) to clusters of three and six users supplied radially by one central heater. Note the discrete changes in flow architecture on the ladder to better organization and performance over time

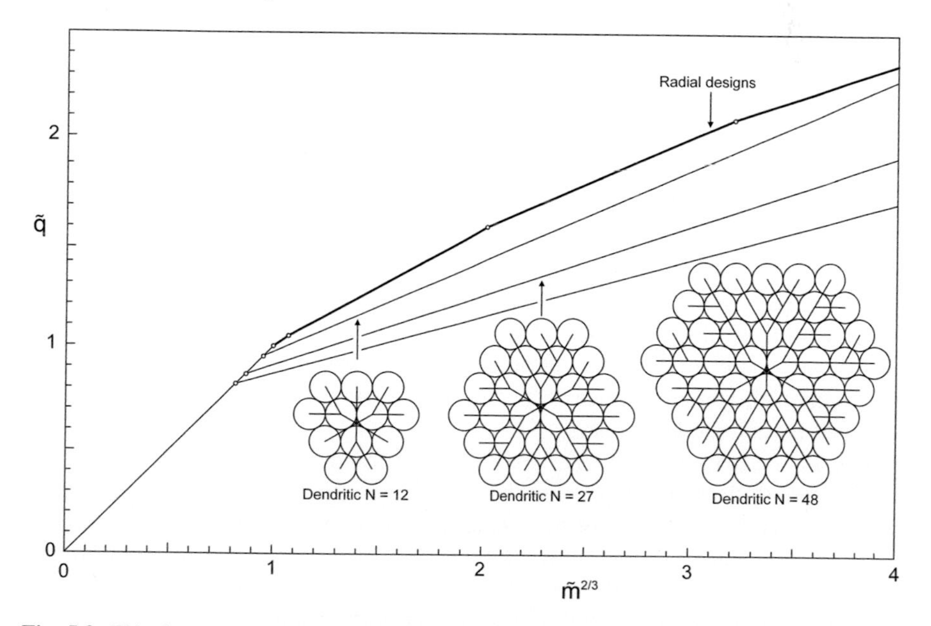

Fig. 5.3 This figure is a continuation of the evolution sketched in Fig. 5.2. Note the beneficial effect of changing the design from radial (Fig. 5.2) to dendritic

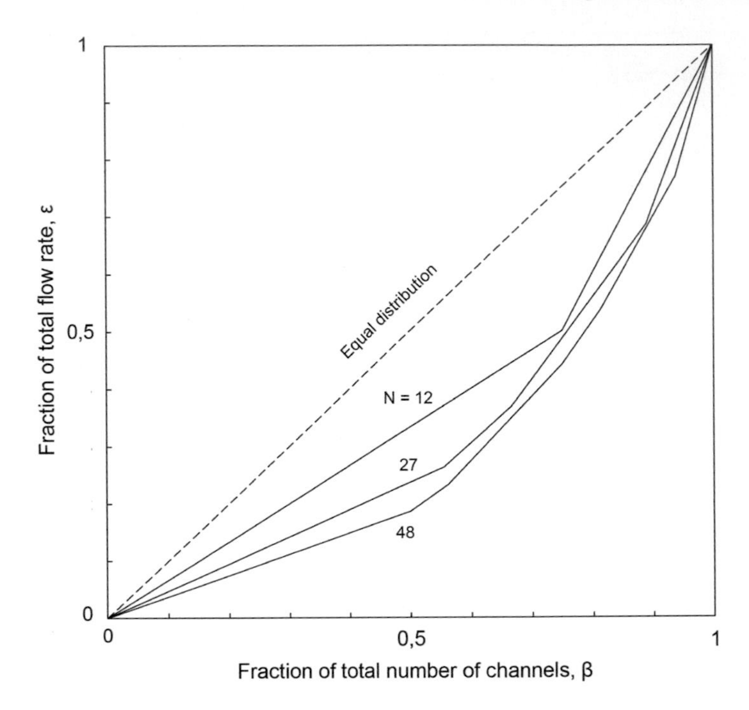

Fig. 5.4 The hierarchical distribution of hot water flow over the population, and the effect of the increasing group size *N*. Note the similar features of this figure and Fig. 5.1

discern by reading Fig. 5.2 from left to right. The water production and distribution must evolve from uniformity to organization, that is, toward nonuniformity.

Radially fed clusters with 3, 6, 12, and 27 users are shown in Fig. 5.2. Tree-shaped distribution designs are shown in Fig. 5.3 for clusters with 12, 27, and 48 users. Clusters with 12 users are displayed in both figures, and they are different. In Fig. 5.2, the distribution lines are radial, and in Fig. 5.3 they are tree-shaped. The more economical design is the tree-shaped, the hierarchical.

The model used for the evolution of the use of hot water reveals two essential features of the natural occurrence of social organization. The first is the urge to unite, to coalesce around a more efficient producer that serves every member of the group. This is the incipient emergence of hierarchy, one large (the center) and many small but equal around the large one.

The second feature is the emergence of more complex hierarchies with intermediate nodes of distribution between the one large in the center and the many small and equal on the periphery. The more complex hierarchies are described graphically as tree-shaped flows. Their complexity increases over time as the society becomes more advanced, as $\tilde{m}$ increases in time (to the right on the abscissas of Figs. 5.2 and 5.3), and the number of individuals who join the organization increases.

We arrive in this way at Fig. 5.4, which illustrates the nonuniform distribution of the commodity (hot water) over the population. This figure is reminiscent of Fig. 4.7.

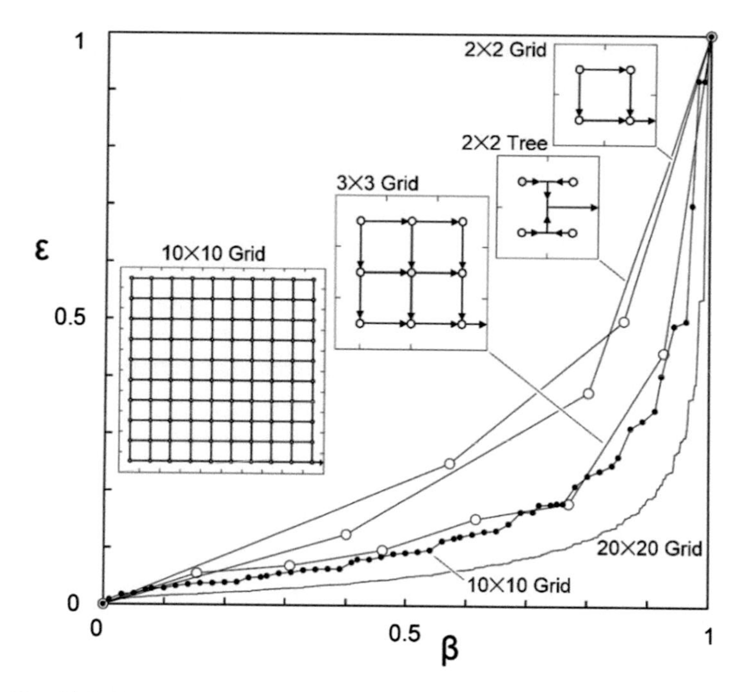

Fig. 5.5 Equal inhabitants, equal inputs, and equal channels are not bathed by equal flow rates on a territory with area-point flow. The fraction of the total flow rate (ε) versus the fraction of the total population (β) on a territory connected by a square grid $n \times n$. The inequality in the distribution of flow over the area becomes more accentuated as the population size increases

The three curves are drawn for the three dendritic designs shown in Fig. 5.3. On the abscissa, from 0 to 1, is the population of flow channels of distribution. On the ordinate is the fraction (ε) of the total flow rate that corresponds to a fraction (β) of the population of channels. The piecewise linear curves are convex, just like in Fig. 5.1 for the river basins. The curves shift downward as the complexity of the organized flow system increases.

This trend agrees with the trend exhibited by river basins, except that the human-made designs have curves (Fig. 5.4) that are located well above the ε–β curves of the river basins (Fig. 5.1). Conclusion: Inequality is considerably less severe when social organization is present.

Inequality happens even when the natural hierarchical channels are destroyed and replaced with an artificial one-size design everywhere, as during communism. To see why, consider the flows presented in Fig. 5.5, because designs more egalitarian than these do not exist. The square territory is inhabited by identical elements, which are connected by identical one-size channels. The center of each element is indicated with a small circle. Each center receives the same flow input as its neighbors (from above, perpendicular to the plane of Fig. 5.5), and that flow is analogous to the uniform rain input M_0 falling on the river basin in Fig. 4.3. The individual inputs collected from all the elements are evacuated as one stream through one corner of the square territory.

The concept of equality, however, is not to be confused with equal individuals and equal links between individuals. In Fig. 5.5, the one-size links form perfectly square grids of $n \times n$ individuals. Three territory sizes are shown for $n = 2$, 3, and 10. For calculating the node-to-node flow rates in cases $n = 3$, 10, and 20, it was assumed that the flow rate is proportional to the node-to-node difference that drives the flow (similar to pressure difference or altitude difference for fluid flow). In the limit $n \rightarrow \infty$, the square territory is bathed by diffusion, with a diffusivity constant (a physical measure) that has the same value over the entire territory, and where the stream evacuated from the whole territory is generated uniformly at every point on the territory, like the rain falling on the plain.

After calculating the individual flow rates between adjacent nodes, we constructed [1] the ε–β curves shown in Fig. 5.5. The flow from the square territory to the outlet is distributed nonuniformly through the uniform grid. Most of the total flow rate ($\varepsilon = 1$) is concentrated in a small group of channels, near $\beta = 1$. Geographically, that group is composed of individuals who happen to be in the vicinity of the point of attraction, which serves as outlet for the big stream.

This is how we discover that inequality persists even when the "one-size" design is imposed artificially. Geography is the reason for inequality in this extreme design. The "equals" who are positioned close to the point source or sink are the huge beneficiaries. This is the physics basis of the birth of oligarchy in post-communism Russia.

The spreading of innovation events over the populated territory is a very subtle way to control inequality. Innovation happens when one individual seizes the opportunity to open the flow channel that he or she controls. This is analogous to opening a valve or flipping a switch for the first time. This local design change attracts more flow into the liberated channel: more flow means that the innovator becomes wealthier (Fig. 4.1). The subtle part is that the flow over the entire territory is also enhanced because of this singular act. The whole population becomes wealthier because of a single innovation. The distribution of wealth over the population becomes more equal than in the absence of innovation (Fig. 5.6).

To illustrate this change and its consequences, consider the flow distributed over the 2×2 grid shown at the top of Fig. 5.5 and also in the lower right corner of Fig. 5.6. This can be analyzed with pencil and paper. When the person-to-person channels are identical, the flows through the channels are distributed nonuniformly. Identical channels mean that the flow resistance is the same in every channel, where "resistance" is the name for the ratio between the node-to-node difference (pressure, voltage, altitude) and the current driven by that difference (fluid flow, electric current, river water).

What happens when one channel opens up? Assume that the resistance of one channel drops to half of the value that it had before this innovation. Figure 5.6 shows two possible locations where this change can happen. In both cases, the ε–β distribution of wealth becomes more equal, and closer to the diagonal of the frame.

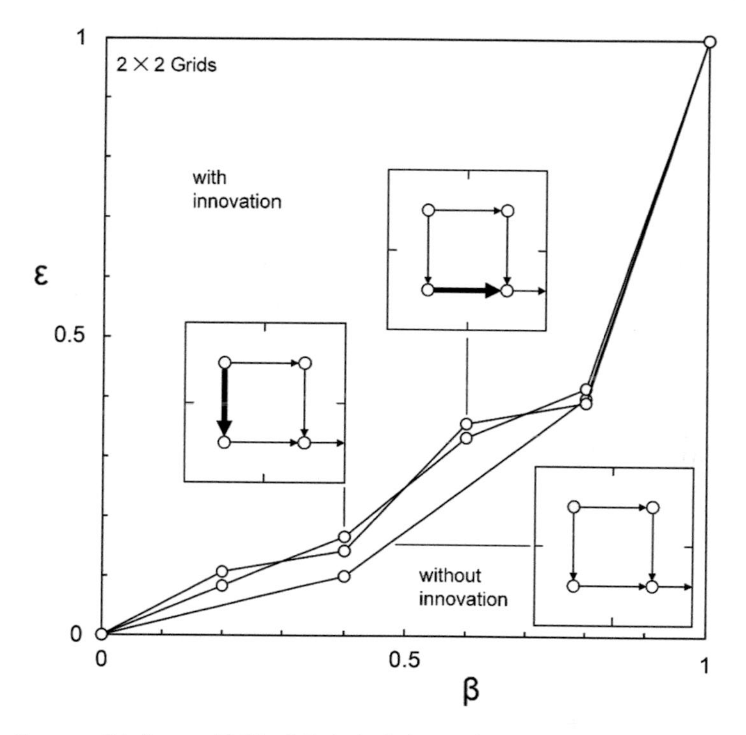

Fig. 5.6 Compare this figure with Fig. 5.5. A single innovation is a local design change that eases the flow and attracts more flow (wealth) to the place of innovation. The result is a more equal distribution of flow over the area

Why, because the innovation triggers an increase in the flow of channels in which the flow rates were not the greatest.

The effect of innovation on controlling inequality is an argument for spreading freedom, education, science, technology, know-how, and especially, the spirit to question and take risks. These add-ons to the design spread the flow to distant patches of the territory that are traditionally not flowing and not known for generating innovation.

In summary, social organization depends upon many factors, many of which are considered intangible. Inequality in movement (or in wealth) is intimately tied to geography, to the physical reality that everything that flows on an inhabited area flows from area to point, or from point to area. Even when the members of the population are equal and equally connected, those who happen to reside near the source or the sink are necessarily visited by a larger stream than the peripheral members. This is the physics basis of organization, and it is evident not only in human social organization but also in animal organization. Organization is born and evolves for the benefit of the whole.

In human society, it is not entirely a matter of happenstance who lives near the "source." Inventions and creative thought tend to occur more in places where they have occurred many times before. This is how the advanced countries, societies,

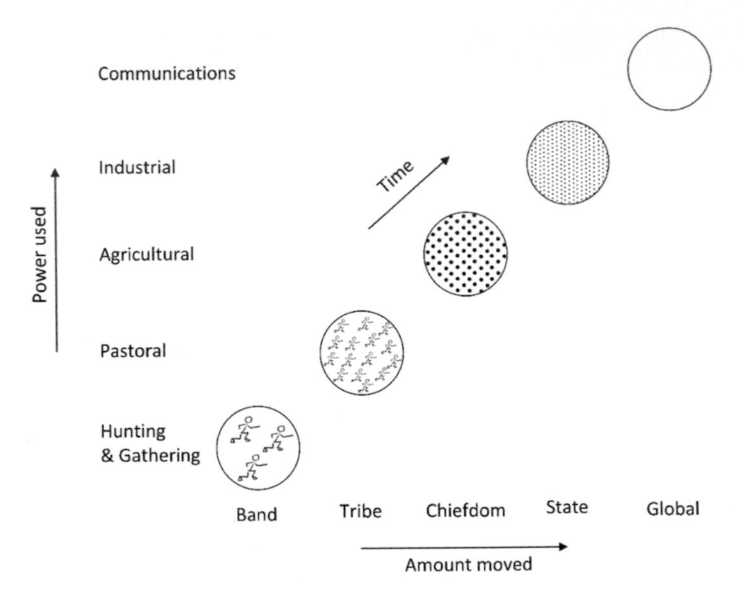

Fig. 5.7 The evolving designs of human social organization are known by more familiar names. In time, the territory swept by organized members increases from the band, tribe, and chiefdom to the state and the globe today. This happened because at the same time the production and use of power increased along with other measures that characterize the sizes of the physical streams driven by the consumed power: subsistence, standard of living, rate of fuel consumption, hot water, traffic, gross domestic product, wealth, affluence, advancement

and territories rose above their neighboring areas. Examples from recent history are presented soon (page 63). The preceding paragraph holds true at the larger scale. The advanced area serves as "source" and attraction for those who are in it or near it. This, by the way, is the physics cause of human migration, and why human migration is unstoppable.

The only way to make the individuals equal and the flows between individuals equal is by distributing the evacuation port uniformly over the territory. In other words, every node on the map would receive or discharge the same flow as its neighbors. This would be a social organization without organization: every hut surviving on its own, not connected to its distant neighbors, as during the hunting and gathering era (Fig. 5.7). That will not happen again, because of the time arrow of evolution (view from left to right the abscissas in Figs. 5.2 and 5.3). Human power, movement, knowledge, and organization have evolved far and irreversibly from what they were even recently, at the dawn of steam power.

Even if sources and channels of a single size were distributed uniformly on the area, the hierarchy and inequality would persist. The efficiency of the whole complex flow architecture is increased over the whole territory when an individual design change (one innovation) is made to liberate the flow locally, in one spot on the flow territory. This is the physics basis for why innovation, which is individual, is of benefit to the whole society. It is addictive, valuable, and indispensable.

The world that we know today comes from individual, local innovations that have enriched inventor and society. The examples are innumerable—here are only a few:

- The invention of steam power in Great Britain, which made Boulton and Watt rich and the whole British empire the richest and most powerful in the 1800s.
- The invention of the distribution of electric power over huge areas (Tesla, Westinghouse), which had an even greater effect than the steam power that generated the electric power.
- The awakening to economic activity in hot and humid tropical and equatorial areas, thanks to the invention of air conditioning by Willis Carrier in 1902.
- The opening of the economy of communist China to the west, which allowed the vigorous and established economic flow architecture of the west to invade the economic wasteland that had been created by communism.
- The never-ending inventions in communications (moving type, bookbinding, typewriter, computer, iPhone), which liberate the flows of society at the point of invention and on the whole territory.

To innovate is to have talent, and this is why innovation and talent have value in society. The physics of value was expressed from the beginnings of society in the word "talent," from the Greek noun *talanton* (Latin *talentum*), which originally meant a scale or balance for weighing a valuable quantity measurable as a weight of gold or silver. In both Greek and Latin, the meaning of talent diversified to include a unit of money, a coin, and this meaning spread throughout the Roman empire and the western civilization that resulted from it.

Innovations, and how they liberate the human flow locally and then globally, offer an unexpected mental viewing of how a society flows and evolves, and where the future of a society resides. Innovations happen almost randomly in space and time, on the map. Each innovation is like a light that comes on and lights up the patch of territory around it. I said "almost randomly" because in every society there are concentrations of people and points of innovation, such as the communities of MIT and Duke on the U.S. map.

The country with more innovators is covered by more flickering lights, like a Christmas tree. The more innovative the society, the brighter its lighted territory, and the more enlightened its people. This is the physics basis of enlightenment [2], and it is the same as the origin of social organization. Innovations are the source of evolution in society. Innovators make history.

Brain design and its flows fit in the same mental image as a society covered and enlightened by flickering innovations that liberate flow. The cortex is covered by an immense number of neurons. Each neuron is connected to tens of thousands of neurons on the cortex. One connection at one point—a new concept, idea, or vision—illuminates the whole cortex, with lights that flash randomly, not synchronically. More numerous and more frequent lights make an enlightened mind. Ephrat Livni [2] summarized this way the unexpected consequence of the physics of social organization:

"When a single individual becomes 'enlightened' by knowledge or a realization, it creates new connections and fortifications in the brain's synapses. A slew of previously unrealized connections may then follow so that the flow of thought is more evolved. The more ideas we encounter and are interested in, the more likely we are to be enlightened thinkers. To put it in physical terms, the metaphorical tree that blocks the river's rush leads a persistent and creative person to an innovation or an insight. They push the tree out of the way somehow, with thought and application. This, in turn, causes more cortical connections and more ideas to flow. That has a good chance of improving not only their lives, but the world we all share."

References

1. A. Bejan, U. Gunes, M.R. Errera, B. Sahin, Social organization: the thermodynamics basis. Int. J. Energy Res. **42**, 3770–3779 (2018)
2. Ephrat Livni, Physics can explain human innovation and enlightenment. *Quartz*, 30 June 2018

Chapter 6
Complexity

Nature is lying naked on the table, and its description—called physics—is concise, simple, unambiguous, and constantly improving. Yet, confusion reigns especially with respect to complexity and other troublesome concepts such as the second law of thermodynamics. In this chapter, we step back and take a look at these notions—at their meaning and definition—on the background provided by nature.

What is discernible all around us is *macroscopic*, not infinitesimal, not statistics, particles, and subparticles. Those who took the reductionist path missed the macroscopic phenomenon, the form, the drawing, the photograph, the sculpture, and the movie. The preceding chapters showed that nature and the phenomena we know best are macroscopic, diverse, multi-scale, complex and morphing, and evolving.

What is complexity? Complexity is a lot like freedom: everybody knows, but very few can tell what it is. Complex is often confused with complicated and random. Here is how to see and tell the difference:

Complexity, organization, and evolution in nature are most powerful and useful when pursued as a discipline, not as jargon. A discipline has precise terms, rules, principles, and usefulness. Why is this important? Let's review the central concepts and words that underpin the physics of evolutionary design today: information, knowledge, evolution, change, arrow of time, pattern, organization, drawings, complexity, fractal dimension, object, icon, model, empiricism, theory, disorder, and second law. Along this path, we will discover that information is not knowledge, fractal dimension is not a measure of complexity, and pattern is not a live flow architecture. Configurations, as physical means to facilitate the flow of knowledge, are subject to the natural tendency toward evolution over time.

During the twentieth century, statistical mechanics, quantum mechanics, information theory, and computer science have changed the scientific discourse on everything, from science itself to what life is. Instead of terms and images that do not require an advanced education, today it seems that legitimacy on this topic comes from speaking "scientific" language, not English. The scientific sounding language is about disorder, uncertainty, scale, emergence, chaos, entropies of innumerable types

A. Bejan, *Freedom and Evolution*,
https://doi.org/10.1007/978-3-030-34009-4_6

and, above all, “information” and information entropy. The fact that very few of us understand this kind of talk is going unnoticed, obviously, because the world does not speak jargon.

> What a delightful thing is the conversation of specialists! One understands nothing and it’s charming. (Edgar Degas)

At age 17, I came home and told my father the veterinarian that the professor who was lecturing on analytical geometry was making me experience headaches because he was all about complicated “general” equations and no drawings. Imagine that, geometry with no drawings. My father said, “do not think about his abstract stuff, simply walk past what makes you unhappy.” His favorite metaphor was the rabid dog: “Act as if the rabid dog does not exist, walk past him, and he will not bite.”

This does not have to continue this way. Let’s not join the crowd, the marching column. Even better, let’s walk against the crowd (Fig. 11.1). Start with the observation that information is not knowledge. The computer is not because it is nothing more than an extension of the human who uses it to move (to live) more easily. It is one artifact among very many. On the other hand, you are, you with that artifact. You are, you make decisions (purposeful choices and changes), and as a consequence you and your group move (live) more easily and with longer lasting power. You are a specimen of the evolving human and machine species. I am another specimen of that evolving species.

Those whose maternal language is not English have no choice but to learn English, and along the way they acquire the habit of checking the dictionary often. That is a huge advantage, and the more the dictionaries of other languages you check, the greater the advantage. Teach your children this: the treasure is in the dictionary, in the meaning of the words, and in their origin. Here is the meaning of some of the key words that do not require an advanced education:

Information is a universal term, like geometry, energy, and physics. It is expressed by the same word in many languages. Information comes from the Latin verb *informo-informare*, which means to give form and shape, to form, and to fashion. In English, it means something told, news, intelligence (as in spying), facts, data, text, and figures. These days, information also means data that can be stored in or retrieved from a computer. This is why ‘information’ sounds abstract, opaque, fancy and high level, as in information theory, information technology, information security, and information age. To see through the jargon, ask your computer what are all those 01 s and 10 s for? They are instructions for how to make drawings, or signs: letters, numerals, text, lines, curves, shades, colors and sounds (musical notes).

In modern languages all over, information means a sign, a signal, such as seeing 24 June instead of 23 June on the headline of today’s newspaper. What you, the observer, decide to do with the sign is you. What I do with it is I. What we do are physical changes, design changes in your movement and mine, with time direction of change. The changes are dynamic, with power, in time. They are the *action* that knowledge is and are not to be confused with mere information.

There was a lot of the said information chiseled on obelisks in Egypt, and it meant absolutely nothing, no message, no imagination, no dream, no design, no change, no action, and no embodiment into anything. It stayed that way for millennia until a young nobody, Jean-François Champollion taught his contemporaries how to hear the sound of words from what was chiseled on obelisks. What you read and do today with Egyptian information is you, not somebody else.

> Having knowledge but lacking the power to express it clearly is no better than never having any ideas at all (Pericles)

Change has a time direction, and its arrow of time is called *evolution*. The ability to affect design change is an integral part of the live system that morphs and evolves in order to move on earth more easily, farther, and longer in time. The "ability" of the moving thing comprises many physical features: freedom to change, power (watts) to move things and reshape flows, access to information, memory of past changes that facilitated movement, memory of changes that are detrimental, and so on, on the staircase to better flow organization over time. This is *action,* and it applies to everything, inanimate and animate. Knowledge is action based on information, not information alone [1].

Here is what information is and how we use it. When I make a drawing or a sequence of drawings (an evolving design) to describe what I see around me, I need to know where to put my lines, how long, how thick, what color, and straight or curved (cf., Chap. 7). To make a drawing I need freedom. The better painter gives himself more freedom.

When I was learning how to draw I was fascinated by how "messy" are the master's pallet and the floor of his atelier. That is no mess, in fact, it is the best niche that the painter built around his working hand and mind so that the painting is his best, and his own pleasure reaches climax. That niche and what the painter puts on canvas are the "information," and they come from imagination, observation, and also from the warehouse of information called education, culture, library, and parents.

In order to get from that warehouse to my drawing on paper, the information must travel, it must flow, and it must be communicated. Today, a lot of that is accomplished digitally, in three steps: from observed object to digital code, from digital code to me, and from my digital information to my drawing of the physical object. Before computers, the same three steps have been in place forever, as everyone can recognize by replacing the word "digital" with eye, brain, book, teacher, education, or training.

Design is a plan, a scheme, a project with purpose, and an intention (an aim) for an outcome. Design is the arrangement of parts, details, form, and color, so as to produce a complete whole that has purpose and performance.

Design is not "pattern." The tree-shaped architecture of so many things that flow (river basin, lung, city traffic) is a changing design because it has purpose (direction in time), which is to facilitate flow between a point and an area, or between a point and a volume. On the contrary, pattern is a static, regular, and an unvarying arrangement of form, parts or elements. The tiles on the bathroom floor and the atoms locked in the crystal lattice have pattern but not design. They are in the "dead state," according to thermodynamics. In them, there is no flow, no change, no morphing, no freedom, no time direction, and therefore no life and evolution.

Optimization means to make a choice, to opt, to select the configuration A as opposed to B, or the rhythm C instead of D. This action, to opt, comes from the Latin verb *opto, optare*, which means to choose, to elect, and to select. It is an expression of the human urge to change things for the better. The direction of the change comes naturally, most often from conscience in humans, and it is the choosing that paves the road to the better. The paved road is evolution.

Optimization is not mathematics. It is not the operation of finding an analytical function that is a simplified description (a facsimile of nature, called "model") of a curved object in nature, then taking the first derivative of that function, setting the derivative equal to zero, and finally solving that equation. This operation is finding the extremum of the function, the hilltop or the valley bottom, and nothing more. It has nothing to do with choosing between two or three concrete options that are useful and available, with the objective of making your own life better.

Organization is a consolidated group of live flowing elements, as in an organ. Organization is a systemized whole (also called "system"), the organs connected and flowing together in the moving animal body, the river channels and wet interstices in the drainage basin, and the components moving together inside the vehicle on the highway. In society, organization is the live group assembled for a specific purpose (activity, movement), such as a club, union, political party, executive structure in business, sports team, university, or government. Design is living organization, not dead pattern.

A picture is worth a thousand words. This is why all these notions, from information (sign, form) to design and organization, occur as images in the mind before they are spoken as words. This is also why the most important keyword to understand is *the image*, which happens in the mind and later comes from the hand as a drawing.

Drawing is an image that can be discerned by the eye and understood by the mind. "Understood" means that the image is connected to other images in the brain in ways that make the storage of images more compact and their retrieval more rapid. The concept of drawing comes from *disegno* in Italian, and *dessin* in French, from which "design" in English. A drawing has three essential features, which are self-standing, independent of each other. The maker of a drawing can attest to the fact that one feature is chosen independently of the others [2]:

(i) The drawing has *size*. Large or small, the size is represented by the length scale of its frame, sheet of paper, canvas, mural, or computer screen. The maker of the drawing selects the size of the image.

(ii) The drawing has a *meaning* (a message) that is conveyed to the viewers. The message is "knowledge" when the recipients *act* and then make changes based on the received message. The message spreads naturally, from those who know to those who need to know. The meaning is represented by one or more features: shapes, structures, aspect ratios (proportions), and the *organization* of all these features on the viewed plane.

Each feature is distinct. The round shape of the cross section of the blood capillary is not to be confused with the bifurcation of the blood vessel, or the pairing of smaller blood vessels into a larger vessel. All the features are

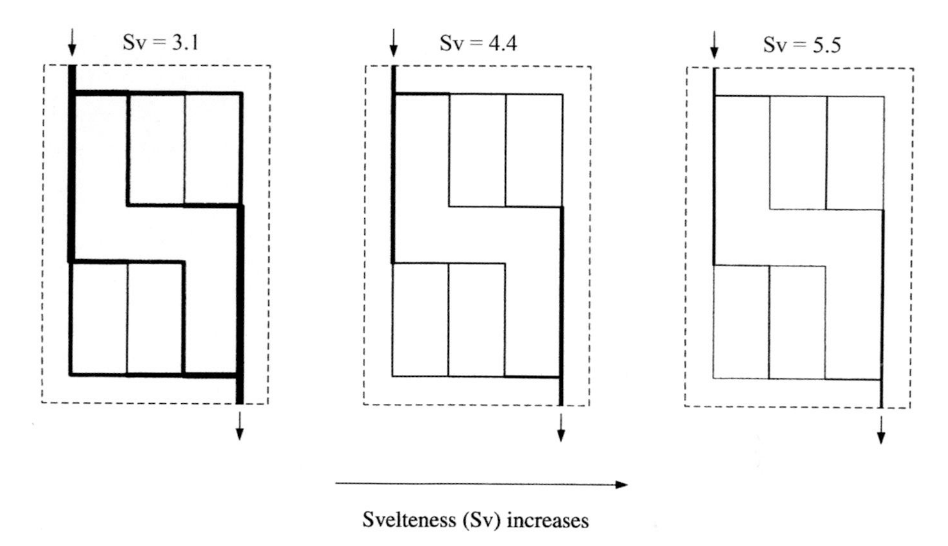

Fig. 6.1 The three drawings are the same, but their lines are progressively thinner. The svelteness property *Sv* of a complex architecture accounts for the thinness of its lines: as svelteness increases, line thicknesses decrease, the drawing becomes sharper and lighter, and it breathes more. The message of the drawing does not change, but its "weight" changes. The svelteness of the flow architecture in this figure was calculated by taking the square root of the area of the dashed rectangle as external length scale, and the square root of the area covered by the channels (shown in black) as internal length scale

organized in a particular way, and as a consequence they convey the message sharply, unequivocally. A simple line drawing of your face today displays the same organization as a drawing made 20 years ago, but a few features are different. The drawing morphs with your age, but the message remains: the portrait is yours.

(iii) The drawing has *svelteness*, which is a measure of the relative thinness of the lines used to convey the message (ii). The svelteness is a dimensionless number defined as

$$Sv = \frac{\text{the external length scale of the drawing}}{\text{the internal length scale of the drawing}}$$

Scale means size in an order of magnitude sense, for example, a length comparable with one meter, not with one millimeter. The drawing made with relatively thin lines has a large *Sv* value (Fig. 6.1). It is svelte looking. The same drawing made with a thicker pen or brush, or copied on a poor copy machine has a smaller *Sv* value. It is heavy looking. A watercolor rendition of the original line drawing has an even smaller *Sv*. A good forgery has a different *Sv* than the original. The *Sv* value belongs to the artist, to one brush and one style of brush strokes, and distinguishes the original artist from the forgerer.

Fig. 6.2 Above a certain height, all turbulent plumes have round cross sections: flat plume rising from a row of smokestacks; round plume rising from a concentrated fire; and plume above a brush fire

Amazing, even the discussion of a single drawing is complicated! Then again, what is complexity?

Complexity is a difficult concept, like chaos, or turbulence. In the beginning, when people knew a lot less than what we know today, complexity meant difficulty, fuzziness, headache, and why bother. Its Latin origin betrays this feeling of defeat: "complex" (*cum* + *plex*, i.e., twisted forms together) comes from the same Latin observation as "perplex" (through + twisted), which also speaks of defeat.

As science progressed, people began to see organization and message in the said complexity. As the thinking became sharper, higher, and deeper, the organization of complexity gave birth to theory, which is the mental viewing with which to predict the observed complexity. Once understood, complexity becomes easier, and we call it architecture, weave, tissue, design, organization, and many more names that are a lot less puzzling.

It happened this way with turbulence, which as a science evolved from fuzziness in the late 1800s (complications of fluid flow were effaced intentionally through a time-averaged description, thanks to Reynolds [3]), to the "large-scale structure" of turbulence in the 1970s, and to the *evolution* of the structure of turbulence, which is now predictable from the constructal law [4]. For example, a flat jet or plume always evolves into a stream with round cross section [5], Fig. 6.2. The reverse is not true: round jets and plumes do not evolve into streams with flat cross sections. This holds true for turbulent and laminar jets and plumes as well.

It will happen the same way with complexity. The language of complexity will be replaced by geometrically precise notions such as size (i), organization (ii), svelteness (iii), and I am sure additional fundamental notions.

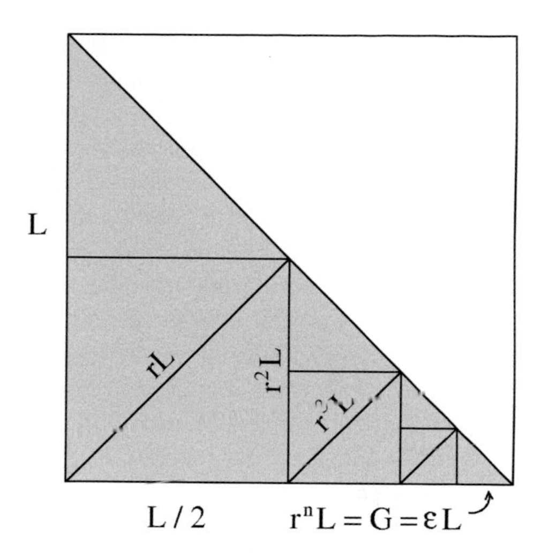

Fig. 6.3 The length of the toothy line increases as its smallest detail (G) becomes smaller

We often read that the fractal dimension (D) of an object (or drawing) is important because it increases with the complexity of the figure, and that it accounts quantitatively for complexity. Had this been true, we would have seen by now a fractal-dimension ranking of drawings according to their complexity, because newly calculated D values appear in the literature unabated. Whether the fractal dimension adds anything to the quantitative description of complexity is questionable.

Consider the drawing made in Fig. 6.3. An isosceles triangle is divided in half by the bisector of its 90° angle. The construction is repeated n times. The bisectors form a toothy line to which we add the left side (L) of the original triangle. The total length of the toothy line is $L_n = L + rL + r^2L + \cdots + r^nL = L(1 - rG/L)/(1-r)$ where $r = 2^{-1/2} = 0.707$, and G is the smallest length segment in the drawing, $G = r^nL$. In the limit $n \rightarrow \infty$, the total length approaches $L_\infty = L/(1-r)$. The actual total length L_n can be expressed in dimensionless form as $\tilde{L}_n = 1 - r\varepsilon$, where $\tilde{L}_n = L_n/L_\infty$, and $\varepsilon = G/L < 1$.

The fractal dimension (D) of the toothy line is defined as $\tilde{L}_n = \varepsilon^{1-D}$, where D depends on ε and r. In the limit $\varepsilon \rightarrow 0$, or $n \rightarrow \infty$, D approaches the irrational number $D_\infty = 1 + r/\ln 10 = 1.3071$. As ε decreases (or as n increases), the D value approaches D_∞ from above. For example, when $n = 2$, ε is 0.5 and $D \cong 1.379$. When $n = 4$, the D value drops to 1.338, which is still above D_∞.

What does all this mean? It means that the drawing of Fig. 6.3 is a "fractal object" strictly in the limit $\varepsilon \rightarrow 0$ where the construction algorithm would have been repeated an infinite number of times, and the number of lines used in making the drawing would be infinite. To the artist who attempts to make the drawing, this would mean that the fractal object would have *infinite* complexity. This, by the way, is the physical reason why the fractal object is impossible to draw and see, and why the "fractal objects" that populate the literature are not fractal. None of them. They do not exist in nature. They are all Euclidean (this according to Mandelbrot himself [6], p. 39), because the algorithm assumed in making the drawing is intentionally and arbitrarily stopped (cut off) at a small length scale that is sufficiently macroscopic so that the drawing can be made, printed, viewed, discerned, remembered, and discussed.

Review now the fractal dimension calculated during the construction of Fig. 6.3. The drawing with infinite complexity has the fractal dimension $D_\infty = 1.3071$, which is finite, not infinite. Second, the Euclidean drawings with finite complexity (finite n and ε) have D values greater than D_∞. All the drawings that the reader can see in Fig. 6.3 are decidedly less complex than the fractal drawing; yet, their calculated dimension D is greater than the fractal dimension D_∞ of the drawing with infinite complexity. When D decreases, the complexity increases.

The fractal object that is talked about in the fractal geometry literature is not an object. The word "object" means a thing that can be seen or touched, or a person or thing to which action, thought, or feeling is directed. The original Latin word, *objectus*, means something thrown in front of you, a thing that appears in front of you (from the verb *objicere*, where *ob* means toward, for, before, and *jacere* means to throw, from which the word "jet" in all Latin based languages and English).

The fractal object is, at best, a thought "in the limit," never achievable, never palpable, and never to be seen, like Sadi Carnot's reversible heat engine. It is not part of nature. Yet, there is a big difference between the two thoughts in the limit: the reversible engine springs in the mind because of physics (the laws of thermodynamics), whereas the fractal object is as arbitrary as the algorithm chosen by the mathematics artist, or the paint and brush chosen by the painter.

> A mathematician may say anything he pleases, but a physicist must be at least partially sane
> (Josiah Willard Gibbs)

Icon is a simple drawing that conveys the same message as numerous and much more complicated drawings. Making drawings is a means of communication that evolved under certain rules, which constitute the discipline of graphic design. The rules are much stricter when the icon is about human safety. For example, the design of the pedestrian "Walk" and "Don't Walk" sign on street corners (Fig. 6.4) must meet traffic regulations. These signs evolved with the technology available for communicating messages visually.

Today, the world has adopted images of a walking person (pictogram) and a raised hand (ideogram) or standing person (pictogram), respectively, to indicate when to cross the street and when not to cross. The raised hand is an icon because it represents the hand gesture of a traffic policeman. These signs have a characteristic size: not too big, not too small. There is a size range that helps most pedestrians in most circumstances (Fig. 6.5). The larger size is easier to see, but it is more expensive to make. The smaller sign is cheaper but it is more difficult to see and understand.

The golden-ratio shape of the sign plays an important role on the speed with which the message is scanned by the two human eyes [7]. The scanning is faster when the image is shaped as a golden-ratio rectangle, like the cinema screen, the computer screen, and the business card. Because the two eyes are aligned horizontally (like the observed world), the scanning of the image is faster by 50 percent horizontally than vertically. For the same reason, the person who is blind in one eye perceives

Fig. 6.4 Evolution from realism to simplicity in the icons that convey the messages "Walk" and "Don't walk." Note that each sign is a golden-ratio rectangle, that is, a rectangle the two sides of which form a ratio comparable to 3:2 and 2:1, or the golden ratio 1.618

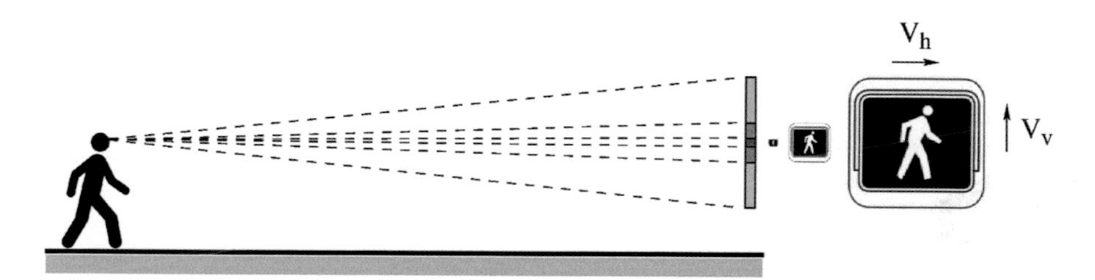

Fig. 6.5 Size possibilities of the "Walk" sign, and the scanning range of the viewer. The V_h and V_v are the speeds with which the two human eyes scan horizontally and vertically the rectangular image, and $V_h > V_v$

faster the image that is rounder, for example the square, not the 3:2 rectangle. This physics holds equally for all animals, two-eyed and one-eyed, because they hold for humans. If we can learn from experiments on animals about humans, we can learn on humans about the animals.

People express a preference for what comes easy. This is a physical psychological phenomenon, present and impossible to disregard. Niederhoffer [8] discovered it in the natural clustering of stock prices. Stock market decision-makers place their limit and stop orders at round numbers with which they are accustomed to deal. Niederhoffer called these numbers "constructal." In them, we see the human preference for round numbers, because they are easier to discern, easier to remember, and faster to tell other people.

The icon represents an individual in the most general sense (male, female, tall, short, old, young). The drawing is an organization of simple elements that resemble the head, torso, legs, and arms (Fig. 6.6). By itself, each element is meaningless.

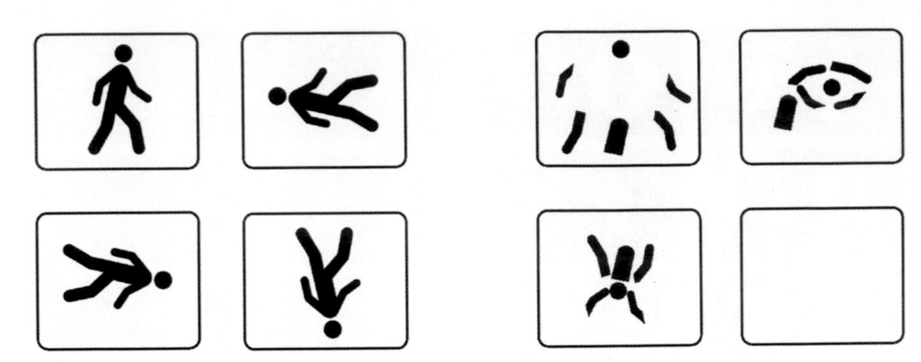

Fig. 6.6 Possible organizations of the shape elements that form the icon "Walk"

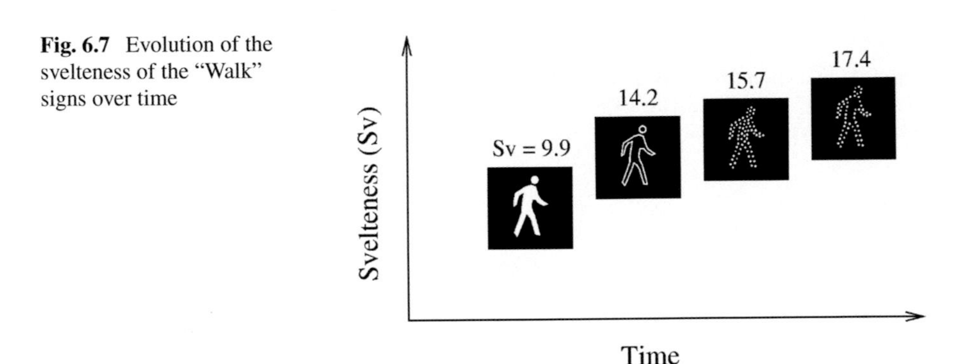

Fig. 6.7 Evolution of the svelteness of the "Walk" signs over time

The meaning is their organization. If the elements are organized in other ways, the pedestrian is puzzled, and the message does not flow from the sign to the mind of the viewer.

Svelteness is an essential property of all drawings, including the icon. In Fig. 6.7, the svelteness of each icon was calculated with the *Sv* definition (p. 69) by taking the effective outline as the internal length scale, and the square root of the lit area (white) as the external length scale. Icons with higher svelteness are newer designs. The icon with the highest svelteness appeals to the knowledge of the pedestrian, that is, it conveys information as a set of lit dots, and the mind of the pedestrian creates the image and then interprets it.

A photograph or a moving picture of a person crossing the street would be more realistic than a line drawing, but would it be more effective? Key is the minimum detail that is sufficient to convey a quick, clear, and safe message. The few and simple lines, or the gestalt effect (the form) of aligned dots that are lit, are thick but not too thick to convey the message to the persons across the street.

Models share one characteristic with icons, their simplicity. Yet, models and icons are different. The icon is a simple drawing of a mental viewing, observed or imagined. The model is strictly about the observed: it is a manmade simplified facsimile of an object or phenomenon observed in nature. The duck from the woodshop is the model,

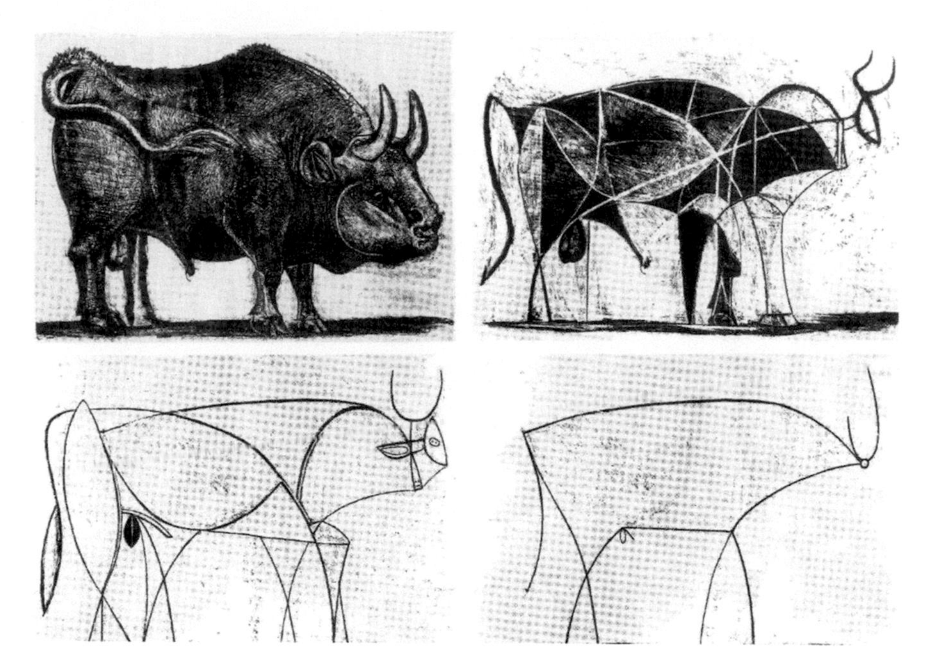

Fig. 6.8 Evolution from realism to simplicity in Picasso's lithographs "The Bull", showing the artist's quest for the essential drawing of the bull

and the duck on the lake is the observed natural object. The human action of modeling is empiricism, which means observation first and description later. Duck first, wood model later. Modeling is the opposite of theory (idea first, comparison with nature later). Modeling is not theory.

Evolution means changes that occur in a discernible direction in time. Evolution, the word, is defined unambiguously at its origin, the Latin verb *evolvo*, *evolvĕre*, which means to roll out, to roll forth. Contrary to today's discourse, evolution is a much older and more encompassing concept of physics (of everything) than Darwinian biology.

Why is simpler better? Drawings—their veracity and complexity—evolve along with the human ability and technology to describe the mental image. The evolution toward the simplest drawing that still captures the message was illustrated by Picasso in the sequence of lithographs called "The Bull". In 11 plates, the artist captured the essence of the message that was ultimately expressed in a small set of organized lines with high svelteness (Fig. 6.8).

Evolution is also evident in the accuracy with which a drawing is made by computer, as technology evolves. Figure 6.9 illustrates this with two circles drawn with pens of two thicknesses. The circles are not perfect, but this is not why they are used as examples. They are used because their svelteness can be calculated as $Sv = p/A_b^{1/2}$, where p is the inner perimeter of the black line, and A_b is the area occupied by the black line. To calculate p and A_b, the circle drawings were digitized with several

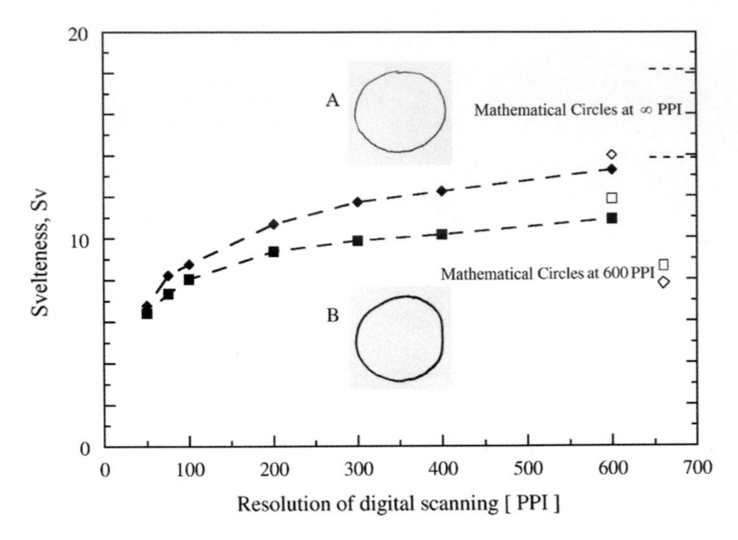

Fig. 6.9 The evolution of the svelteness of two hand-drawn circles, as the scanning resolution (points per inch, PPI) increases

resolution settings measured as PPI (points per inch), which in Fig. 6.9 are plotted on the abscissa. The scanned drawings were converted into binary (black and white) images in order to calculate their p and A_b values.

Figure 6.9 shows that Sv is not a constant, unlike the Sv of a mathematical circle with rim of constant thickness. The svelteness of the hand-drawn circle increases monotonically as the scanning and reproduction technology improves toward more PPI. The increasing trend is due to the rough and ill-defined edges between the black trace left by the pen and the white paper. The texture of the paper and the force on the pen on paper are features that belong to the particular hand that makes the drawing: they cannot be reproduced fully, not even in the limit of infinite machine power. Each curve (Sv versus PPI) is like the length of the coast of Britain, which was the calculation that served as starting point for fractal geometry [6].

Disorder is a concept that deserves particular questioning. Along with the claim that nature evolves toward increasing complexity, we often read that the natural tendency is toward greater disorder, and that this tendency is commanded by the second law of thermodynamics. This is wrong, as one can see by reading the statement of the second law [9–11] (also in Chap. 1, p. 8).

We often read that the second law states that "entropy must increase," and that the "classical" laws of thermodynamics pertain to "equilibrium states." Many even teach that thermodynamics should be called thermo "statics." Such statements are not thermodynamics. I collected several false statements [9, 10], from which I quote one, from a physicist: "the second law applies to closed macroscopic systems consisting of an extremely large number of particles, such as liquids or gases." This is not true. The second law statements hold for *any system*, open, closed, isolated, adiabatic, steady state, unsteady state, with configuration, without configuration, and evolving or not.

The second law says nothing about disorder. Many confuse the second law with the view that in a box filled with imaginary identical particles the assembly tends toward a larger number of possible energy states [12, 13]. This is the core idea of statistical mechanics, not thermodynamics, yet lost in the teaching of it are three important observations:

First, to assume a swarm of identical particles in a closed box is to throw away the "any system" power of thermodynamics. The "any system" is the most general system in physics. It is the system with unspecified organization and with infinite freedom. Compared with it, the box with bouncing particles is a very special case, with presumed internal components and rigid impermeable boundary.

Second, particles and disorder are not common observations. They are concepts, not phenomena. From such language, how can there be a "law of increasing disorder"? This has been a source of confusion in science because as we look around we are struck by the complete opposite phenomenon: design, change after change (evolution), self-organization, emergence, and order out of what some would describe as lack of order.

Third, decades before statistical mechanics, the second law and the first law were stated with reference to systems of unspecified size (e.g., heat engines), not infinitesimal. Engines are flow architectures that are macroscopic, organized, and evolutionary. Order, not disorder, is their chief characteristic and claim to fame. Engines are every day futuristic, not "classical." They are full of life and motion, not in "equilibrium." They are eminently dynamic, not "static."

Phenomenon is the human observation that certain things happen innumerable times the same way. One phenomenon represents one natural tendency, which is distinct from other natural tendencies. To observe and describe a phenomenon is *empiricism*.

Law of physics is a compact statement (text, or formula) that summarizes one phenomenon.

Theory is to rely on the law to experience a purely mental viewing of how something *should be*.

The phenomenon covered by the first law of thermodynamics was known in mechanics as "what goes up must come down," and also as *vis viva* (live force) and *vis mortua* (dead force). Today, we recognize this more generally as the conservation of energy, from kinetic to potential when a body is thrown upward, to the energy flow (from heat into work) through a closed system such as a heat engine operating in cycles.

The phenomenon covered by the second law is the one-way tendency of all flows, such as the flow of water under the bridge, and the flow of heat from hot to cold. Today, we recognize this natural tendency as *irreversibility*. Every flow, by itself, proceeds from high to low. Fluid flows through a duct from high pressure to low pressure. Heat through an insulation leaks from high temperature to low temperature. If you do not know beforehand which is the high and which is the low, then the direction of the flow will tell you. Why, because it is the law, and any system in nature obeys the law.

The phenomena observed as complexity, organization, design, and the other terms reviewed in this chapter are natural organization, evolution, and life [14, 15]. The occurrence and evolution of freely morphing configurations is present in everything that flows and moves more easily over time. This phenomenon is represented by the constructal law. Observations of this kind are everywhere: river basin evolution, lung architecture evolution, city traffic evolution, heat exchanger evolution, and aircraft evolution. These observations reveal the arrow of time [16] in nature, which points from existing flow configurations to new configurations through which the flowing is easier. Not the other way around. Why, because this is the law, and all systems in nature obey the law.

Words matter, especially in science. This is why it was essential to define unambiguously the terms of any discussion about complexity. When you hear somebody conflate disorder with the second law, ask that speaker to state the second law. You will then discover that the speaker does not know the subject even though before your question the speaker was convinced that he did. For you, the discovery is much more precious: you will suddenly see in front of you the empty suit and, if the speaker is well established, no suit at all, just the new clothes of the emperor.

The birth of the word "constructal" is a beautiful story of why words matter, and why it pays to be curious while still ignorant, unbiased, but creative. I had discovered the principle in September 1995, and I was using it to predict and draw flow architectures (heat flow, fluid flow, pedestrians). The principle had no name. One of my doctoral students at the time, Marcelo Errera from Brazil, came into my office and said that my drawings look like the drawings in a book called "The Fractal Geometry of Nature". I was unaware of that book and fractal geometry. I asked him to get the book from the Duke library and show it to me. When I opened it, on one of the first pages, Mandelbrot (born and raised as a child in Poland) explained that he invented the word "fractal" from the Latin verb *frangere* (accent on the "ah"), which means to break something into two smaller pieces. I immediately realized that he was wrong about the Latin: the word "frangere" is preserved 100% in Romanian, in spelling and pronunciation, and obviously nature does not "snap" sticks and bones into smaller bits all the way to dust. The time arrow of design evolution in nature is the complete opposite of that.

I told Marcelo that Mandelbrot should have used the opposite of frangere, which is the verb *construĕre* (*construire* in Romanian, *construer* in Portuguese), and we laughed that "the geometry of nature is constructal, not fractal."

By chance, at the time I was revising two manuscripts right before publication, and I inserted the word "constructal" and its Latin meaning in those articles [17, 18], both published in 1996. To my surprise, the readers decided that constructal was a useful word. Perhaps you also had this experience: something that you create for pure fun is later taken as serious, heavy, and profound by scientists steeped in deep contemplation.

In this story, helpful was the fact that Marcelo and I had been constantly comparing Portuguese with Romanian, Italian, Spanish, French, and Latin. We had fun with stories about the origin of languages, people, migrations, ideas, and words. I still collaborate with Marcelo, who is now a professor in Brazil.

Fractal algorithms are descriptive, not predictive. Fractal geometry is no theory. One guesses an algorithm that would lead to a drawing that resembles a natural image. The fractal mathematician is no fool: he or she does not reveal to you the multitude of algorithms that led to drawings that looked like nothing. The masters were no fools either: they did not show anybody the discarded paintings strewn all over the floor.

The constructal law is predictive. It teaches how to discover the drawing (and the "fractal algorithm," if you wish) and how to predict the evolution—the morphing—of the natural configuration over time. Description is empiricism, and it is very common, banal, diverse, and abundant. Prediction requires a law and a theory based on that law. Prediction is extremely rare because it unifies the abundant phenomena.

Science needs both: the many small and the few large, the diversity and the unifying view, the many descriptions and the extremely few predictions, the abundant empiricism and the rare theory.

References

1. A. Bejan, *The Physics of Life: The Evolution of Everything* (St. Martin's Press, New York, 2016)
2. A. Bejan, M.R. Errera, Complexity, organization, evolution, and constructal law. J. Appl. Phys. **119**, article 074901 (2016)
3. O. Reynolds, An experimental investigation of the circumstances which determine the motion of water in parallel channels shall be direct or sinuous and of the law of resistance in parallel channels. Philos. Trans. R. Soc. **174**, 935–982 (1883)
4. A. Bejan, *Convection Heat Transfer*, 4th ed. (Wiley, Hoboken, 2013), chapters 6–9
5. A. Bejan, S. Ziaei, S. Lorente, Evolution: why all plumes and jets evolve to round cross sections. Sci. Rep. **4**, 4730 (2014)
6. B. Mandelbrot, *The Fractal Geometry of Nature* (Freeman, New York, 1982)
7. A. Bejan, The Golden Ratio predicted: vision, cognition and locomotion as a single design in nature. Int. J. Des. Nat. Ecodynamics **4**(2), 97–104 (2009)
8. V. Niederhoffer, Clustering of stock prices. Oper. Res. **13**(2), 258–265 (1965)
9. A. Bejan, Evolution in thermodynamics. Appl. Phys. Rev. **4**, 011305 (2017)
10. A. Bejan, Thermodynamics today. Energy **160**, 1208–1219 (2018)
11. A. Bejan, *Advanced Engineering Thermodynamics*, 4th edn. (Wiley, Hoboken, 2016)
12. D.F. Styer, Insight into entropy. Am. J. Phys. **68**(12), 1090–1096 (2000)
13. F.L. Lambert, Disorder—a cracked crutch for supporting entropy discussions. J Chem Ed **79**(2), 187–192 (2002)
14. T. Basak, The law of life: The bridge between Physics and Biology. Phys. Life Rev. **8**, 249–252 (2011)
15. A.F. Miguel, The physics principle of the generation of flow configuration. Phys Life Rev **8**, 243–244 (2011)

16. A. Bejan, Maxwell's demons everywhere: evolving design as the arrow of time. Sci. Rep. **4**(4017), 10 (2014). https://doi.org/10.1038/srep04017
17. A. Bejan, Street network theory of organization in nature. J. Adv. Transp. **30**(2), 85–107 (1996)
18. A. Bejan, Constructal-theory network of conducting paths for cooling a heat generating volume. Int. J. Heat Mass Transf. **40**, (4), 799–816 (1997); published on 1 November 1996, not in 1997, as shown in Fig. 3.3 in A. Bejan, Technology evolution, from the constructal law, in E.M. Sparrow, Y.I. Cho, J.P. Abraham, eds., *Advances in Heat Transfer* (Academic Press, Burlington, 2013), pp. 183–207

Chapter 7
Discipline

The preceding chapters unveiled the physics basis for phenomena and concepts that traditionally are not acknowledged in physics: freedom to change, evolution, design, complexity, life, performance, economies of scale, diminishing returns, innovation, and social organization. These concepts constitute a new body of knowledge.

When the knowledge develops faster than the language, the new is often described in ad hoc terms that vary from one author to the next. In time, the language catches up by zeroing in on the fewest new terms and rules that govern the new domain. This becomes the new discipline. When this state is reached, the new body of knowledge is more powerful, more applicable, and easier to pass on to new generations.

Well-known examples of disciplines are geometry, mechanics, and thermodynamics. The noun discipline is not to be confused with "disciplined," although the two words apply to the scientist who knows the disciplines, the most crispy fundamentals, not the noise. As we will see, there is no contradiction between freedom and reliance on discipline. On the contrary, the scientist who possesses the disciplines is the most free to venture into new territories of knowledge. Why, because that scientist's confidence is justified by the well-tested reliability of the disciplines.

This chapter continues where the previous left off. To find its discipline, the science of complexity can benefit from the example set by thermodynamics, which is a discipline with unambiguous words, rules, and principles. Changing the meaning of the words in mid-course to benefit a narrative, opinion, or political objective is not allowed. Complexity and, more generally, organization and evolution in nature are most powerful and useful when pursued as a discipline.

The science of form began with geometry. I take the reader back to the birth of geometry 2,500 years ago, when the rules for expressing your reasoning (your argument that you are right) by making drawings were extremely few and very precise. While arguing, you were allowed to use nothing more than a straight edge (a ruler) and a compass. The straight edge was for drawing straight lines. The compass was for recording the distance between two points, and for drawing arcs of circles.

A. Bejan, *Freedom and Evolution*,
https://doi.org/10.1007/978-3-030-34009-4_7

The fewer the rules, the more unforgiving the discipline, and the more lasting the reasoning, the proof of the truth, the theorem, and the image (the drawing). No hand-waving allowed, just the graphic construction with two tools available to even the poorest participant in the discussion (homework problem: how could the poorest person build a compass?).

Drawing cannot be displaced by photography. You can draw what you see and what appears in your mind in the dark of the night. You can photograph only what you see. That's the difference between man and machine, and why new machines (artifacts such as AI) cannot displace the human and machine species.

Compare the straight edge and compass with the wealth of drafting instruments available in modern times: templates of many shapes and sizes, triangular, elliptical, French curves, protractors, right angles, 60° angles, 30° angles, and drafting tables fitted with mechanisms that enable the straight edge to sweep the table while remaining parallel to itself. None of these is allowed in the discipline of proving truths with drawings. These tools were not available when the discipline was born.

As an aside, a reader asked, "Does anyone use these instruments any more in the age of the computer?" That's like asking does anyone use paint, brush, and pencil anymore in the age of photography? Sure, the old instruments are still used by those who hear the calling of their imagination and creativity. You are a lot more likely to know a painter who can take photographs than a photographer who can paint or draw. You are a lot more likely to find a graphic artist who can make computer drawings than a computer user who can draw by hand. Furthermore, you can be certain that the computer user is totally unaware of the principles and disciplines that underpin the graphics software of the computer: geometry, perspective, analytical geometry, and descriptive geometry.

Compared with the advanced person who uses modern equipment, the thinker with straight edge and compass may strike us as backward, or prehistoric. This impression is false and damaging to education. The person who learns to use the ancient tools is equipping himself or herself with discipline: how to think most simply and directly, how to deliver the *coup de grâce* while arguing scientifically, how to disarm the skeptic, and how to design even better modern instruments for the graphics needed today.

The thinker who returns to the minimum number of rules of the discipline runs against the fashion that pervades through high school and higher education, the legacy of which is today's knowledge "industry." In mathematics, reliance on the pocket calculator for everything from addition to calculating an integral has done irreparable damage. In thermodynamics, reliance on software packages has created the illusion that the teaching is now more effective, when in fact very few learn the discipline that underpins the phenomena, the laws, and the software of thermodynamics.

Begin with Fig. 7.1, and ask yourself how to divide a segment in half while using nothing but a ruler and a compass. Here is how, and I hope you enjoy the power of the simple:

Use the ruler to draw the segment $\overline{AB}$. With the compass needle at A, draw the arc of a circle with radius a little smaller than $\overline{AB}$. Move the compass needle to B, and draw a second arc with the same radius as the first arc. The two arcs intersect

Fig. 7.1 How to divide a segment in half, and how to draw a 90° angle, while using nothing but a ruler and a compass

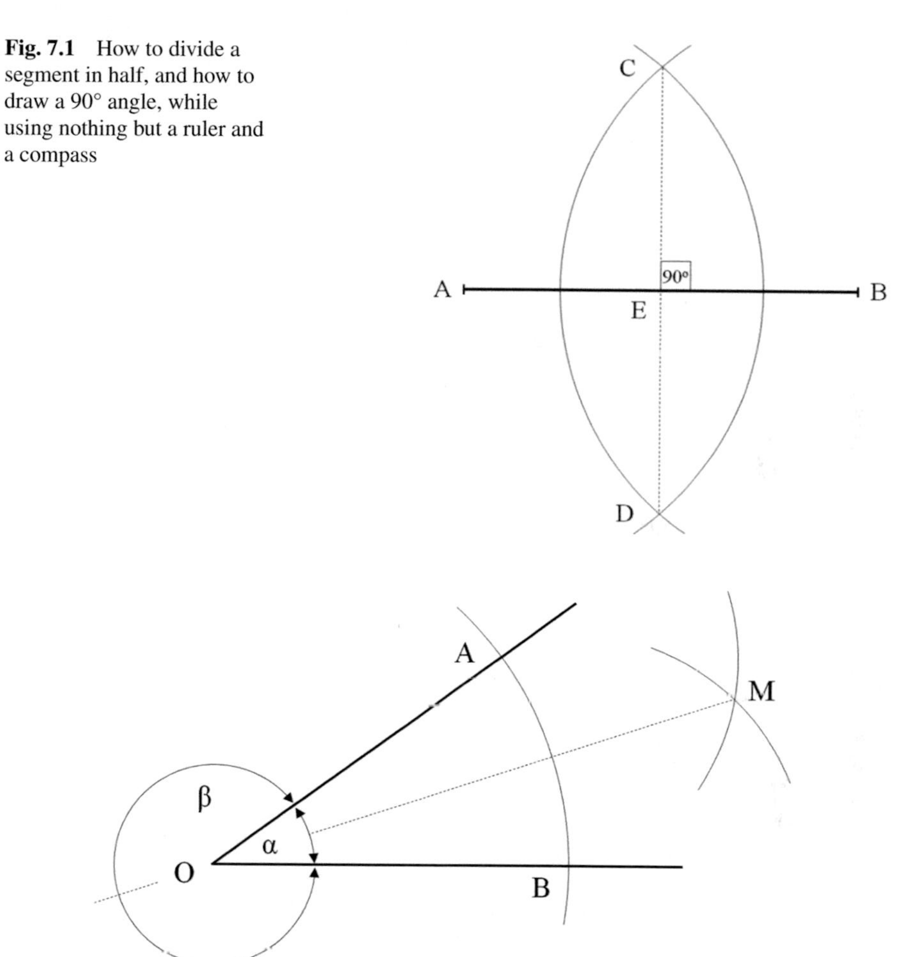

Fig. 7.2 How to divide an angle in half

at two points, C and D, which define the line of symmetry of the construction. The segment $\overline{CD}$ is perpendicular to the segment $\overline{AB}$, and it divides the segment in half. Use the ruler to connect C with D, and discover that $\overline{CD}$ cuts the segment $\overline{AB}$ exactly in half, at point E.

An additional feature that results from this construction is the right angle, which is evident at the intersection between CD and AB. Often, the right angle is symbolized as a small square at the intersection—a square, because the square fits snuggly in that corner.

How about angles that are fractions of angles that we know how to construct, for example, the right angle? Figure 7.2 shows how to divide any angle in half, which means to find its bisector and then draw it. First, consider the acute angle α, where the cuts A and B are made by drawing an arbitrary circle centered at O. Next, with

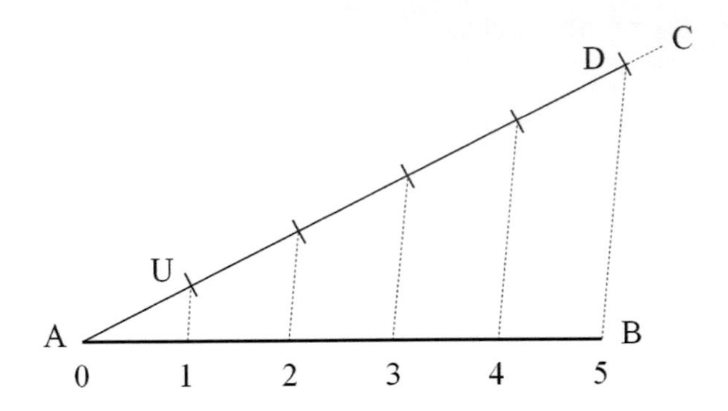

Fig. 7.3 How to divide a segment into any number of equal segments

the compass centered at A, draw an arc with a radius comparable with the distance from A to B. Do the same with the compass at B. The two arcs intersect at the point M, which is equally far from the two straight lines that define the given angle. The line OM is the bisector of α. This construction amounts to discovering the new angle $\alpha/2$.

By repeating the construction of Fig. 7.2, we have the ability to discover all the subsequent angle fractions ($\alpha/4$, $\alpha/8$, …) by using only a ruler and a compass. If the given angle is obtuse, for example, β in Fig. 7.2, the construction is the same as for α, because the bisector of the angle α is the same as the bisector of the angle $\beta = 2\pi - \alpha$.

So far, the constructions have been about dividing segments and angles in half. How about fractions smaller than 1/2? Fig. 7.3 shows how to divide a given segment $\overline{AB}$ into any number of equal segments, for example, five. Draw the arbitrary line $\overline{AC}$ at an angle comparable with 45° relative to the given segment. With the compass, mark the arbitrary distance $\overline{AU}$ on this auxiliary line, and repeat marking this distance five times in the direction from A to C. The fifth mark is the point D. Connect D with B, and recognize the triangle ADB, where the side $\overline{AD}$ has been cut by you in five equal segments. From the remaining marks such as U, draw the lines that are parallel to $\overline{DB}$, and label with 1, 2, 3, and 4 their intersections with the base segment $\overline{AB}$. The outcome is that the original segment is cut into five equal segments, because any triangle such as AU1 is similar to the biggest triangle, ADB.

The segment $\overline{01}$ is exactly one-fifth of the distance from A to B because you have cut your auxiliary segment $\overline{AD}$ into exactly five equal segments. The given line was "cut," which is why the original name for the "cut line" in Latin is *linaea abscissa*. This construction and its name are the source of the term "abscissa" for the segmented lines that frame routinely the graphs in scientific publications.

If in Fig. 7.3 we mark only two segments of size $\overline{AU}$ in the direction from A to C, the construction yields only one cut on the base segment, and that cut is halfway between A and B. This special case is an alternative to the construction made in Fig. 7.1, and for this reason Fig. 7.3 is a generalization of Fig. 7.1.

To divide angles in fractions smaller than 1/2, we proceed on a case-by-case basis. This is not a disadvantage; on the contrary, it is an opportunity to think, to view in space, and to develop dexterity. It's like practicing dribbling and shooting, alone in

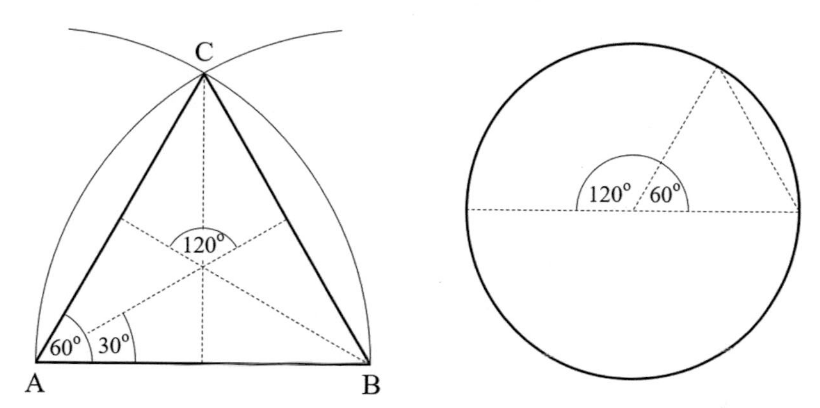

Fig. 7.4 How to draw the 120° and 60° angles, or how to divide a circle into three and six equal arcs

the arena, at night. For example, start with dividing a circle into three equal arcs, each facing an angle of 120° at the center. The construction is the same as inscribing an equilateral triangle in that circle. The equilateral triangle is constructed as shown in Fig. 7.4, the left side. Select the segment AB as the side of the equilateral triangle, and then draw two arcs of radius AB with the centers at A and B. The arcs intersect at C, and the construction assures us that the three sides of the resulting triangle have the same length.

This construction yields two angles, the 60° angle between two sides that touch, and the double of that, which is the 120° angle visible now at the intersection of the bisectors of two of the 60° angles. Delivered by this construction is also the 30° angle, which follows from the 60° angle by using the construction shown in Fig. 7.2. The circle that would be circumscribed to the equilateral triangle turns out to be divided into three equal arcs.

What if the construction must start with the given circle, not with the triangle? In such a case, we first construct the equilateral triangle shown on the right side of Fig. 7.4, which is a repeat of the construction showing on the left side. Alternatively, we measure the 60° angle from the left side of Fig. 7.4 and export that measurement to the circle on the right side of Fig. 7.4. Note the equilateral triangle drawn with dashed line. Two such triangles, adjacent at the center, define the 120° angle that serves to divide the circle into three equal arcs.

How can we export an angle—any angle—from one drawing to another? The generally applicable construction is shown in the upper half of Fig. 7.5. The given angle is α. On the left side, we make two measurements of our own choosing. We draw the circle of radius $\overline{OA}$ centered at O, and then with the compass we measure the radius $\overline{AB}$ of a second circle centered at A. With these measurements, $\overline{OA}$ and $\overline{AB}$, we move over to the new drawing on the right side. We draw the first circle, we mark the point A on the horizontal line, and then we draw the second circle. The intersection of the two circles is point B, which (connected with the tip O) completes the construction of the angle α on the new drawing.

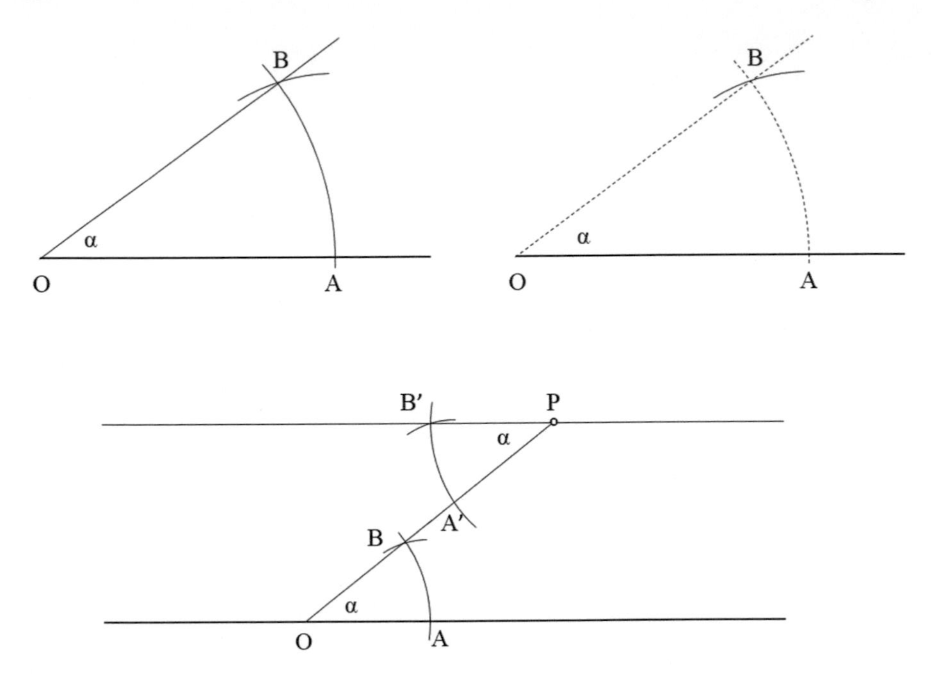

Fig. 7.5 How to measure an angle (α) of an existing drawing (left), to transfer it to a new drawing (right)

Measuring an angle and drawing it somewhere else is the generator of other elemental constructions. For example, the lower half of Fig. 7.5 shows how to draw through any point (P) a line that is parallel to a given line that does not pass through P. The solution is to draw the arbitrarily oblique line PO, which forms the angle α with the original line. Next, we measure α by constructing the triangle AOB, and then we export the angle construction to the top of the oblique line, where P is now the tip of the reproduced angle α. There are only two (both arbitrary) circle radii in this construction, i.e., two measurements, $OA = PA'$ and $AB = A'B'$. This is how we determine the position of B′. The line drawn through P and the new point B′ is parallel to the original line, because the original angle α and its reproduction are equal. They are called alternate interior angles.

These simple examples are just the opening pages to an entire history of everything (geometry, architecture, construction) that preceded the modern tools and edifices that are available today. The ruler and the compass are like the ball and the goalposts: while the shirts, shoes, balls, and stadiums become better and more expensive, the game remains the same.

With practice, the audacity to construct more complicated figures grows. Examine, for example, my India ink drawing in Fig. 7.6, and overlook its small details [1]. Its main characteristic is that I made it with just a ruler and a compass, which, by the way, were my tools as an MIT student in 1969 and after. Note that this drawing is

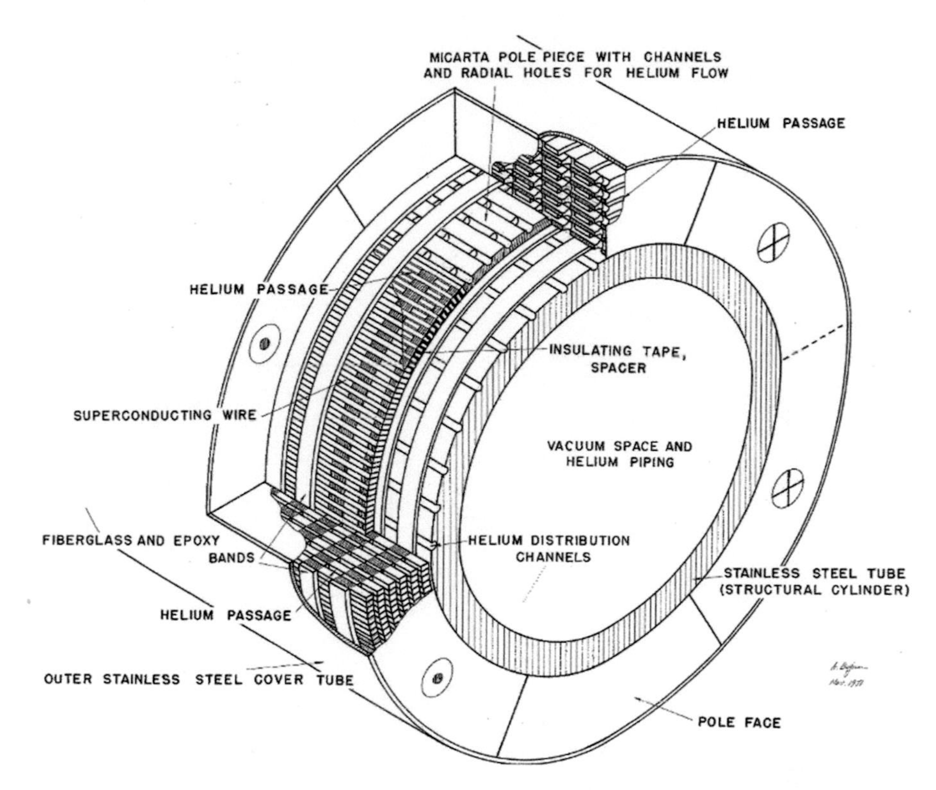

Fig. 7.6 The three-dimensional architecture of electrical conductors and liquid helium channels in the long section of the superconducting winding mounted on the rotor of an electric power generator. Helium flows in all directions, longitudinally, radially, and circumferentially

dominated by numerous "concentric" ellipses of four specified sizes. How can one draw even two nested ellipses? How can one draw just a single ellipse?

The ellipse does not have one center. It has two focal points (two fires, *foci* in Latin, the plural of *focus*, which means fire). The true elliptical shape was drawn by computer at the top of Fig. 7.7, and its two foci are not shown. The horizontal segment that would be drawn through the two foci is the long axis of the ellipse. The vertical segment that passes through the middle of the long axis (as in the construction of Fig. 7.1) is the short axis.

Key is the observation that the long axis cuts the ellipse at its equatorial latitude where the radius of curvature of the ellipse is small. The short axis cuts the ellipse at the poles, where the radius of curvature of the elliptical shape is large. Two radii of curvature, one small and the other large, are the hint that the elliptical shape can be drawn approximately with just one compass by making two kinds of circles, one smaller than the other.

The bottom half of Fig. 7.7 shows how. First, on a blank page draw vertically the short axis, from A to B. With the compass needle at A, draw the large arc of radius $\overline{AB}$. Repeat this with the compass needle at B. The drawing begins to look like the

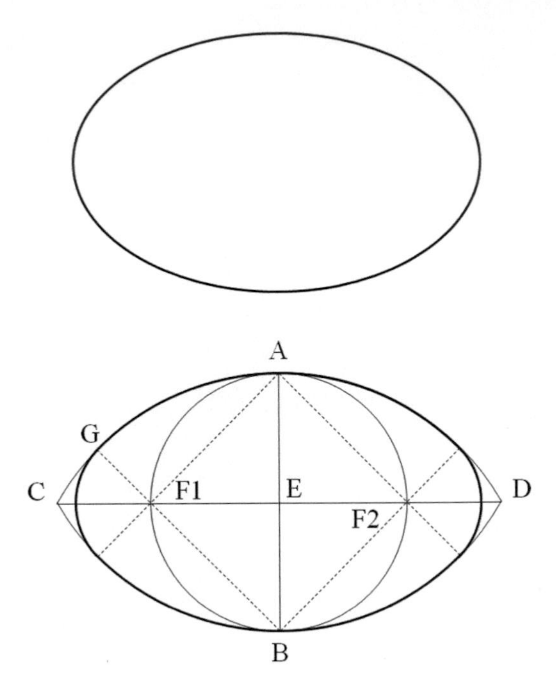

Fig. 7.7 How to draw an ellipse with a ruler and a compass. The upper ellipse is exact, the lower is approximate. Can you see a difference between the two?

human eye, or like the ball used in American football. The two large arcs intersect at C and D. The line $\overline{CD}$ cuts the short axis in half, at point E, cf. Fig. 7.1. This line is collinear with the long axis of the approximate ellipse that we are constructing.

The next challenge is to find the approximate locations of the two foci. These would have to be to the left and right of E. Possible locations are the equatorial points (F1, F2) obtained by intersecting $\overline{CD}$ with the circle of diameter $\overline{AB}$.

The auxiliary lines connecting A with F1 and B with F1 are perpendicular to each other, and, consequently, the small circle of radius $\overline{F1G}$ is tangent at G to the big circle of radius $\overline{BG}$, which is equal to $\overline{AB}$. The two small arcs centered at F1 and F2 complete the construction that I used multiple times in making Fig. 7.6. The resulting ellipse is smooth because of the continuity of slope at the points of tangency, G. The shape (the slenderness) of this particular ellipse is such that the long axis exceeds by 59% the short axis.

Once learned, this construction can be used with dexterity to express in three dimensions what words can never explain. Figure 7.8 shows my India ink drawing of three-dimensional (curved) surfaces in rolling contact at one point, which is the invisible origin of the x-y-z frame [2]. Making this image with my "secret" knowledge and prehistoric tools was fun, warm memories of my earliest education. I get the biggest kick today when colleagues and students who see such drawings assume immediately that they were made by computer.

The evolving designs of nature are surprisingly organized such that their essential features can be summarized by geometric constructions that relate them to the size of the flowing system. With the constructal law, such formulas are predicted, not

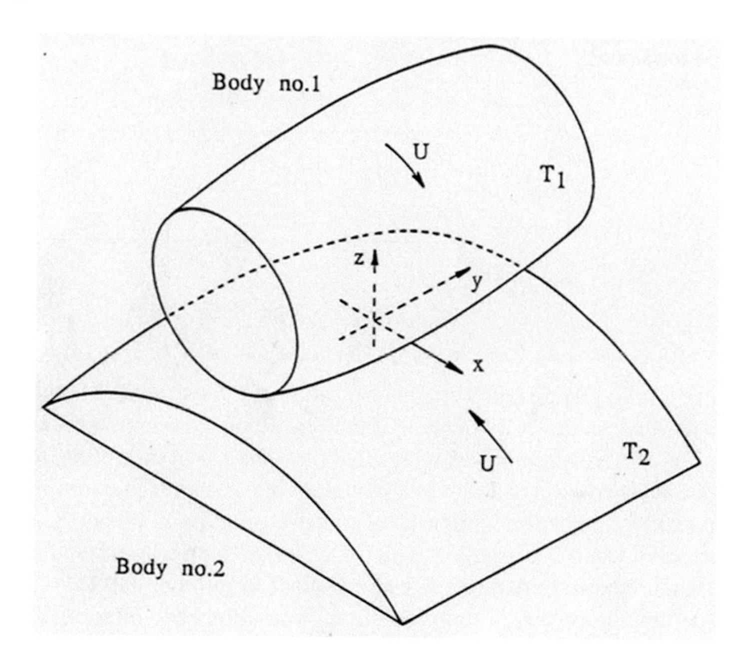

Fig. 7.8 Ellipses, in a drawing of two convex surfaces (two rollers) touching at one point

observed. Predicted formulas are not models. They are theory, not empiricism [3, 4]. Each formula is called a "power law" because it has the form $Y = aX^b$, where Y is the essential feature of the flow architecture (for example, animal speed, animal organ, or airplane engine size), and X is the body size (volume, mass, weight, body length). The factor a and the exponent b are such that the formula matches approximately the (X, Y) measurements of the architectures that abound in nature.

The formula $Y = aX^b$ is a power law because the argument X is raised to a power, b. Many authors mistakenly call $Y = aX^b$ an "exponential," because of the exponent b. The mistake is that in an exponential function Y(X) the argument X appears in the exponent, this way: $Y = a \cdot b^X$.

With few exceptions, the exponent b in $Y = aX^b$ is less than 1, and the corresponding X versus Y relation is named *allometric*. One example is $b = 1/6$, which was predicted for the relations between animal speed (flying, running, swimming) and body mass [5]. Fewer are the relations with $b = 1$, as the relation between jet engine size and airplane weight, and the relation between wingspan and fuselage length [6]. The power law with $b = 1$ is called *isometric* (which means equal measure, from Greek), because doubling X is reflected in the doubling of Y. On the other hand, when the exponent b is smaller than 1, the doubling of X leads to a smaller increase in Y. This is why power law relations with b smaller than 1 are called allometric (from *allo*, which in Greek means departure from the normal). We may regard the isometric relations as special cases of the more general class, which is allometric.

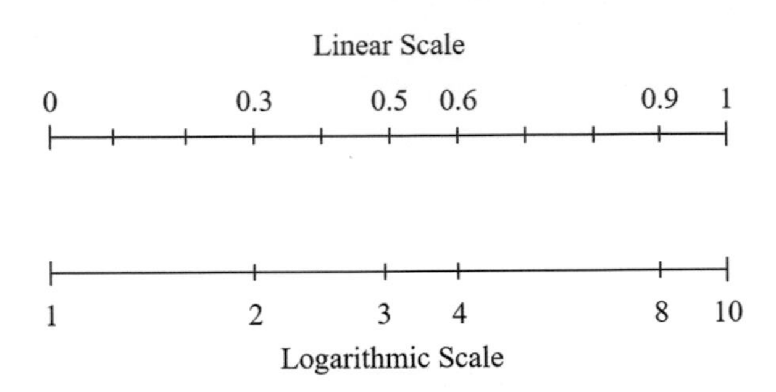

Fig. 7.9 How to construct by hand your own logarithmic scale

Why am I telling you this, in a chapter on geometric constructions, other than for the big reason that nature is a weave of freely evolving flow architectures that are represented by relatively simple drawings that form their own discipline, the science of form? The subtle reason is that when plotted in linear cartesian coordinates (with equidistant ticks in both directions) the allometric relation Y versus X is a curve, which is concave when b is smaller than 1. Curves are harder to draw than straight lines and even harder to remember as being distinct from other curves.

Consequently, the preferred representation of an allometric relation $Y = aX^b$ is in a cartesian frame where the tics along X and Y correspond to log(X) and log(Y). This representation is more useful because on a cartesian plot with log(X) on the abscissa and log(Y) on the ordinate the allometric relation becomes linear, namely, $\log(Y) = \log(a) + b \log(X)$, and is represented by a straight line. The slope of the line is b, and log(a) is a constant.

Lines on log–log graphs are easy to draw, compare, and remember. Before making such a graph, however, one must have at his disposal an abscissa and an ordinate with the tics scribed according to log(X) and log(Y), not X and Y. One must find some log–log graph paper. This is easy today because log–log paper is available in modern bookstores along with the other tools for the knowledge industry. The construction I teach next is how to scribe your own log–log axes when you do not have access to the store, and when you are a professor who likes to draw log–log axes correctly with chalk on the blackboard.

The technique consists of a slight modification of Fig. 7.3. First, in Fig. 7.9 you divide the horizontal segment [the future log(X) abscissa] into ten equal segments: if the abscissa were linear, the tics would represent the values 0, 0.1, 0.2, …, 1. The linearly cut line is shown in the upper part of Fig. 7.9. Next, check the values of the logarithms of the integers from 1 to 10, and you make these startling observations:

- The logarithm of 1 is zero; therefore, the left extremity of the logarithmic abscissa is labeled 1.
- The logarithm of 10 in base 10 is 1; therefore, the right extremity of the logarithmic abscissa is labeled 10.
- The log of 3 is 0.48; therefore, the middle of your new logarithmic abscissa corresponds approximately to 3.
- The log of 4 is 0.6; therefore, under the 0.6 of the linear abscissa, you write 4 on the logarithmic abscissa.

- The log of 8 is 0.9; therefore, under the 0.9 of the linear abscissa, you write 8 on the logarithmic version.
- The mark for log(2) is exactly halfway between 1 and 4 on the log scale. Note that on the logarithmic scale the marks 1, 2, 4, and 8 are equidistant (homework problem: why?).

With all these tics on the new abscissa, it is easy to guess where the missing tics would fall. For example, 2 will fall between 1 and 3, closer to 3. Furthermore, 5 will fall to the right of 4, and the distance from 4 to 5 must be shorter than from 3 to 4. The tics become denser near the right end of the segment, and the denser they are the less important it is to distinguish between adjacent tics, such as between log(8) and log(9).

Now we begin to see why the discipline of geometric constructions that predate computers are essential, empowering, and liberating, especially for academics who are also scientists. This old knowledge helps us see the correct drawing in our imagination. It also helps us make a correct drawing with chalk on the blackboard, and with pencil on paper. This knowledge also empowers us to call out those scientists who are less than truthful with their graphics.

Today, in courses and textbooks, I teach the philosophy that the science of form is a discipline. I did not discover the discipline—this took shape naturally in the evolving drawings of many scientists, most notably Gaspard Monge (1746–1818), the founder of descriptive geometry. Here is a brief sequence of steps toward design with freedom and discipline [7]:

1. *Define your system*: Identify clearly and unambiguously what "system" you are talking about. What constitutes your system? Start with the definition of "system." The system is the region in space, or the amount of mass, that is the subject of your thinking, analysis, and design. The system is yours. The system belongs to the observer. The system is part of you. Everything else is the environment, the surroundings, the rest of the world.
2. *Identify the flows*: Make sure your system has the freedom to change, and that you understand "what flows" within it, i.e., why your system is a "flow system," and why it has freedom.
3. *Start simple*: Allow only one feature of your system to change at first. This endows your system with one degree of freedom. Study if and how the changes to this feature increase the flow access of the currents that inhabit your system. Incorporate into your design the first feature with which you found that your system performs best (be alert, this is not the end!).
4. *Add a degree of freedom*: Allow a second feature of your system to change freely. As you investigate this second degree of freedom, you will find another "best" feature, and you adopt it. With this second feature in place, go back to step 3 and refine that first feature to work with the second.
5. *And another…*: Allow a third feature to vary freely, find the best variant of this feature, and then go back and repeat steps 3 and 4, i.e., refine the preceding two features.

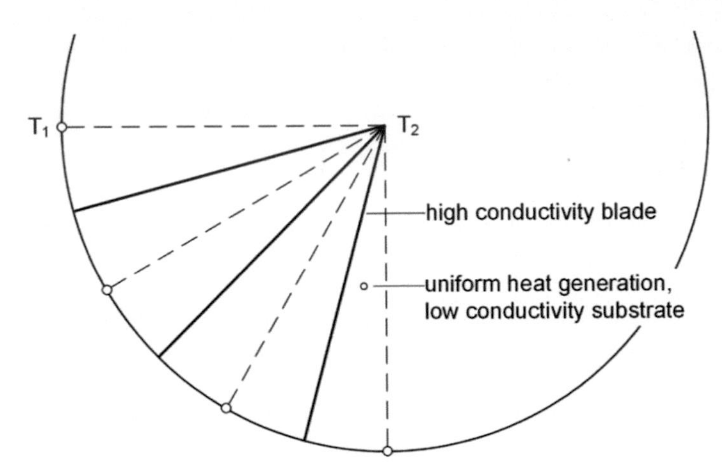

Fig. 7.10 The flow of heat out of a pizza-shaped body is facilitated by inserting several radial blades with high thermal conductivity. The evolutionary design of this heat flow architecture proceeds toward pizza slices that have a certain shape, which means that the whole pizza should have a particular number of slices

6. *And so on*: This is a dynamic construction with no end, except the finite time and life of the thinker.

In the quest for natural architecture, I learned that I have to start modest: leaving just one degree of freedom to the morphing flow architecture is complicated enough. In Fig. 7.10, for example, I decided to seek only the shape of the pizza slice, not its size—that was left for later and to others. Heat is generated uniformly in the disc-shaped body, as in many high-density packages of electronics. Easier heat flow out of the body means lower temperature differences between its hot spots and the heat sink all around the body.

In evolutionary changes toward easier heat flow, the high-conductivity material is more valuable than the low-conductivity materials. This is why the high-conductivity material is expensive, and why only a fraction of the volume can be allocated to it. The object consists of just two materials, in fixed proportions, but these materials are not "mixed." They are unlike two metals in an alloy. The two materials are *organized* such that heat is conducted easier through the whole object. The object has configuration, contrast, and drawing. The less expensive material is the background on which the expensive lines are painted.

In the evolution of technology, the sequence of steps 1–6 happens naturally but slowly, in haphazard bursts of individual creativity. Usually, one step (one degree of freedom) represents a single invention, such as Traian Vuia's use of air-tube tires on the first airplanes more than one century ago. With the new discipline of the physics of form, today industries can fast-forward the design evolution of their technologies and reduce trial and error. In the remaining chapters, we explore the benefits and freedom that come from discipline.

References

1. A. Bejan, *Superconductive field winding for a 2 MVA synchronous generator*, MS thesis, MIT, Cambridge, MA (1972)
2. A. Bejan, Theory of rolling contact heat transfer. J. Heat Trans. **111**, 257–263 (1989)
3. A. Bejan, *Shape and Structure, from Engineering to Nature* (Cambridge University Press, Cambridge, UK, 2000)
4. A. Bejan, J.P. Zane, *Design in Nature* (Doubleday, New York, 2012)
5. A. Bejan, J.H. Marden, Unifying constructal theory for scale effects in running, swimming and flying. J. Exp. Biol. **209**, 238–248 (2006)
6. A. Bejan, J.D. Charles, S. Lorente, The evolution of airplanes. J. Appl. Phys. **116**, 044901 (2014)
7. A.W. Kosner, Freedom is good for design, How to use Constructal Theory to liberate any flow system. Forbes, 18 March 2012 (interview with Adrian Bejan)

Chapter 8
Diversity

The discipline of the science of form also accounts for why objects look similar. In geometry, the concept is called similitude, and it has strict rules. Two triangles of different sizes are similar when the angles are the same in both drawings. All squares are similar, no matter their size. In the discipline of fluid mechanics, similitude is used widely in problem formulation and the graphic display of the flow fields and what the flows carry. The boundary layer regions of laminar flows are geometrically similar. The distributions of velocity and temperature across boundary layers are known for their "similar" profiles, which are delivered as solutions to "similarity" formulations (simplifications) of the Navier–Stokes equations [1].

Nature is dominated by objects and images that hint at similarity but are not similar in the geometric sense. They are dissimilar, or diverse. This aspect is the origin of the saying that every snowflake is unique and every tree is unique, even though all snowflakes and trees look alike because a single principle unites them all [2].

Why so much diversity? Why don't we all look the same? The answer is in the freedom that flow architectures have in how they spread, migrate, and combine with flow designs that they encounter in their paths. In this chapter, I illustrate the birth of diversity from freedom with examples from human evolution, and spreading and empowerment through science and technology.

Prehistoric wood carvings from the Urals 11,500 years ago (from the early Holocene, the current warm period) [3] show that the human face displays a balance between what lies below the eyes and what lies above them. The balance is confirmed by the three faces drawn in Fig. 8.1.

If you are curious, you develop the habit of looking at human faces as you walk through life. If you never leave your village, in Europe, East Asia, or sub-Saharan Africa, you see the same kind of face around you, the kind of face that you drew. In that case, you lose interest in the origin of the balanced features of the face, which may have surprised you once. Look at the faces drawn by artists from a particular region: they look the same, rounder ovals with the eyes at mid-height in Europe and East Asia (Fig. 8.1, right), and more slender ovals with higher eyes in Africa (Fig. 8.1,

A. Bejan, *Freedom and Evolution*,
https://doi.org/10.1007/978-3-030-34009-4_8

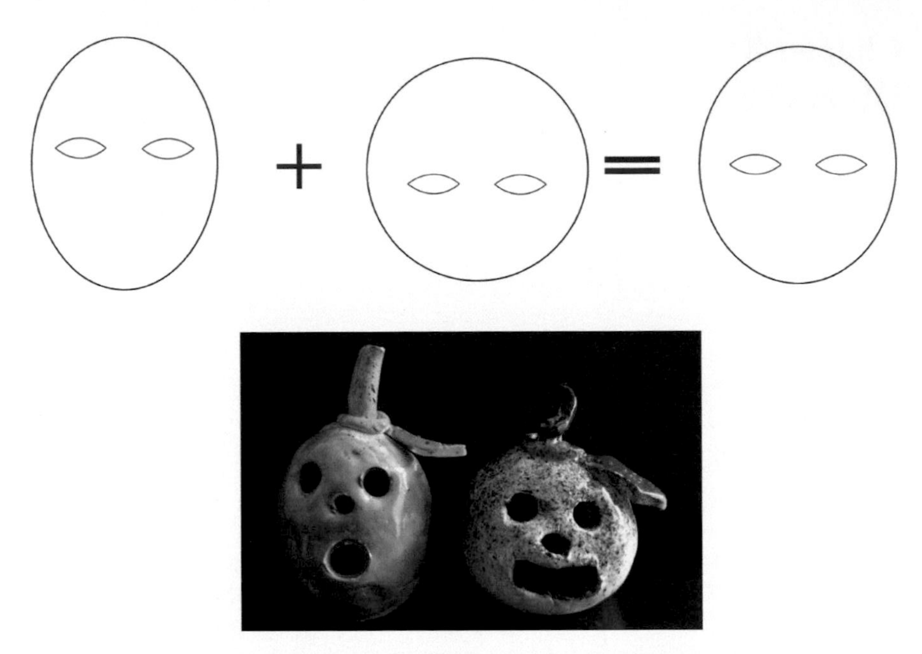

Fig. 8.1 Two human faces, or three? The differences between the first and the third are due to the interbreeding of the first with the second. Bottom: what little children bring home from art class

left). Those fortunate to leave the village and see the world (even by watching movies) notice this difference.

How did this difference happen? What is its origin? Aren't we all specimens of one species, the *Homo sapiens*? Of course we are, but the complete name of that species will emerge later in Figs. 8.3, 8.4, 8.5 and 8.6, as you read this story. The clue lies in the observation that north of the Sahara the faces are proportioned as they are in Europe. The reason for the similarity is the Mediterranean Sea, which from the time of the Neanderthals, through boating, acted as melting pot for the inhabitants of Europe, the Middle East, and North Africa.

The melting pot, the mixing, and interbreeding is the answer. Mixing with whom? With no recourse to anthropology and the human genome, Fig. 8.1 provides the answer. Migrating out of Africa, the *Homo sapiens* interbred with other hominids (Neanderthals) who had rounder faces, lower eyes, and a larger cranial volume. These people are represented by the middle drawing. Who were these people?

My evolution from drawing the ellipse (Fig. 7.7) to questioning human faces took several decades. Questioning the origin of the difference between left and right in Fig. 8.1 came as a flash in 2009, after publishing two constructal theories, one that predicted in retrospect all the animal speeds and frequencies (flying, running, swimming) [4], and the other for predicting the evolution of speeds in athletics, the 100 m sprint and the 100 m freestyle in swimming [5]. Both theories attracted attention as science news on the web and in the press. Stronger was the interest in predicting speeds in athletics, which coincided with the 2008 Beijing Olympics and predicted (from a physics principle, the constructal law) that future winners will emerge from competitors who are bigger and taller.

Wake-up events that change a scientist's orientation are rare, yet they do happen. Why rare, because a scientist who is worth his or her name is already awake, curious, excited, and happy, but stuck in his love tunnel of ideas, questioning and surprising people. Even the most curious and creative thinkers find it extremely difficult to question "what is," which is the obvious that is all around and not changing.

My wake-up call to question my own origin came one month after publishing the evolution of speed records in running and swimming. In August 2009, I received a surprise email from Prof. Edward Jones, a nutritionist who studies obesity and its spread over body shapes that differ by origin, African, European, and East Asian. Prof. Jones is an African American and a former athlete. He asked me why the fastest runners are black and the fastest swimmers are white. I was stunned. My student Jordan Charles and I had predicted the evolutionary future of running and swimming, but never questioned the obvious segregation of the winners according to their origin.

Today, I know why I was blind to the "origin" aspect. I was raised as a performance athlete in the 1960s. That was the golden era of sports—the restart of the Olympics after the Second World War. The Olympics continue today as a healthy form of education due to Pierre de Coubertin, toward creating a "truly egalitarian elite." I was bought up not to question the advantage or disadvantage that my adversary has. That was not even a thought. I wanted to play better and to win, and it meant to train more and harder, to live clean, and to study. My competition was the competition with myself.

By the way, Pierre de Coubertin, descendant of French nobility, hit the nail on the head with his dream of creating a "truly egalitarian elite." He saw no conflict between equality (the *égalité* slogan of the French revolution) and hierarchy—the elite—which is omnipresent, most visibly in athletics, academia, movement on the globe, and wealth.

The answer to Prof. Jones' question came almost right away, because of the body shape data that he showed me. Bodies of sub-Saharan origin have longer legs and arms and shorter torsos than European bodies of the same height. Consequently, the African body has a higher center of gravity (Fig. 8.2, left). This makes the African body run "taller" than the European body of the same size, and this is why the athlete of African origin has a 1.5% speed advantage in the sprint. The longer torso gives the athlete of European origin a 1.5% speed advantage in swimming (Fig. 8.2, right).

We published this physics theory of the "divergent evolution" of speed in sports [6], and our article generated even more positive feedback from the general public, readers, journalists, and scientists. A small fraction of this group was questioning why we would even talk about "race" in sports, and instead was offering "poverty" as explanation for the divergent evolution: lack of swimming pools for black children in America, which is as false as arguing that surfaces for running are not available in "poor" Europe. Others argued that muscle fibers in blacks are different than in whites, and that the body density in blacks is greater than in whites (hardly, all bodies

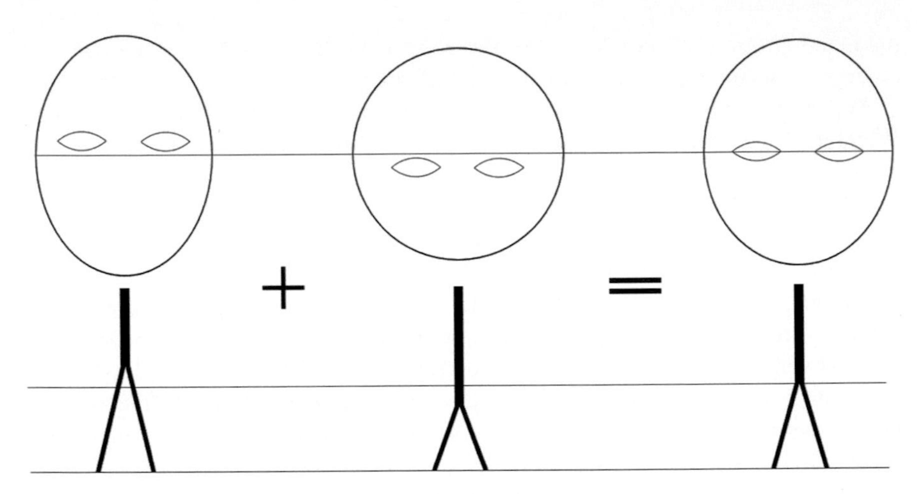

Fig. 8.2 Two human body types: African (left) versus European (right). The differences between the two are due to the interbreeding between the human wave out of Africa with other hominids living east and north of Africa

are essentially water, almost as dense as water; furthermore, what does body density have to do with success in running?).

Even if correct, such arguments are at best descriptive, not predictive with regard to the divergent evolution of the speed sports toward blacks on land *and*, at the same time, whites in water. Others even argued that blacks are the fastest sprinters today because during slavery American blacks were "bred" for strength to work on plantations. Bred by whom, according to what scientific technique? In any case, slavery in the Americas lasted 300–400 years and longer elsewhere, and continues today. It is real, it is to be condemned and eradicated, but on the background of human evolution (Figs. 8.3, 8.4, 8.5 and 8.6) a few centuries are a blip that does not count.

Lost to these critics was why East Asians were not among the fastest runners and swimmers, even though they have all the incentives and support (wealth) to excel in these probes. Just look at the government support for athletes in communist China. The reason why East Asians do not figure in the black–white divergence in speed records is that unlike Africans and Europeans, the East Asians are noticeably shorter. Perhaps, they are a bit different because of the third event of interbreeding between the *Homo sapiens* and other hominids, which distinguishes the East Asians from the Europeans and Middle Easterners who experienced only two interbreeding events (see Fig. 8.3).

Questions from readers about why blacks are faster and jump higher continued to arrive. A student from the UK was writing a graduate-level thesis on possible "environmental reasons" why the human body in Africa "evolved differently" than in Europe. I did not tell him about Fig. 8.3, because that would have been bad teaching. Instead, I told him to question himself where the Europeans come from, and what mixing occurred on the world map as *Homo sapiens* migrated out of Africa.

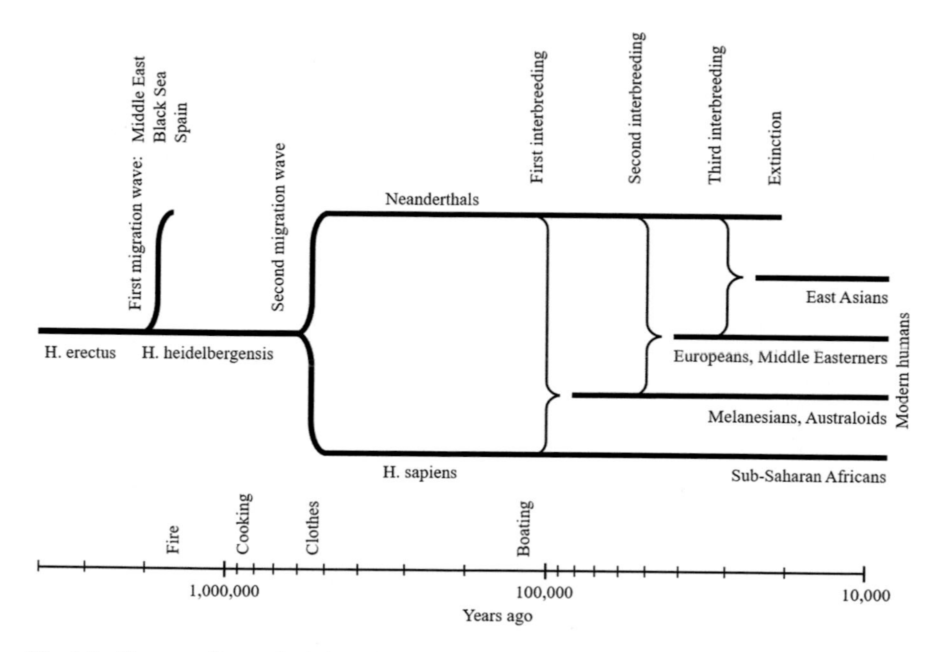

Fig. 8.3 The spreading and mixing of modern humans on the globe. The time scale is logarithmic

I told this student that his question about body types is not about human evolution, because human evolution took place on a very long time scale, in Africa. His question is about a different phenomenon, which is called growth (or spreading) [7], the growth of the area inhabited by humans. Growth is happening on a much shorter time scale. For humans, evolution took millions of years, whereas the growth of one specimen (one lifetime) takes roughly 20 years, as shown in Fig. 8.4.

Growth is not evolution. They are often confused because they both exhibit changes in flow configuration over time. The fundamental difference between the two is why I drew Fig. 8.4 in two dimensions, with time running in both directions, to the right and upward. The difference between the two time scales is so immense that the phenomenon of "growth" makes a cut of zero thickness on the horizontal timeline. In that cut resides all that lived, grew, and died in the population that inhabited the globe at that time. Only two specimens of that population are shown on the vertical, under one adult *Homo sapiens* on the horizontal. There were so many couplings, so many births, and so many encounters with other living things, plagues, and environmental changes, that one cut was followed by another cut on the horizontal axis, on the slow-moving timeline of evolution.

The fundamental difference between growth and evolution is expressed graphically in Fig. 8.4 through the fact that in the horizontal alignment of bodies from ape to modern human each image represents an adult. This is not the case in the vertical direction, where all ages during the life of the specimen represent the phenomenon of growth.

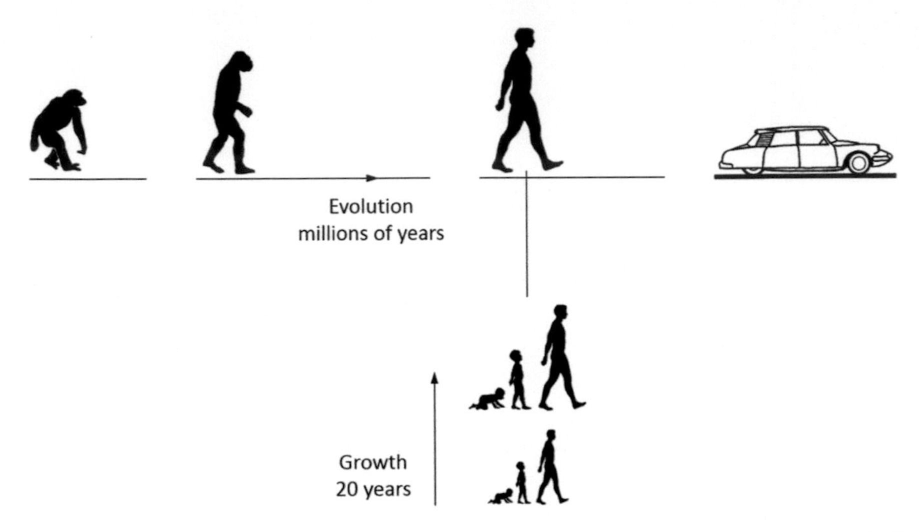

Fig. 8.4 Evolution and growth are two fundamentally different phenomena in nature. The clock ticks in both directions, to the right on the horizontal, and upward on the vertical. The time scales in the two directions are vastly different. In the vertical direction, there is a single species, and each specimen is morphing continuously, from birth to adulthood and death. On the contrary, the horizontal axis is discontinuous, all the specimens are adults, and each species has become extinct, including the naked *Homo sapiens*. The *homo* species that rules the world now and in the near future is the *human and machine* species

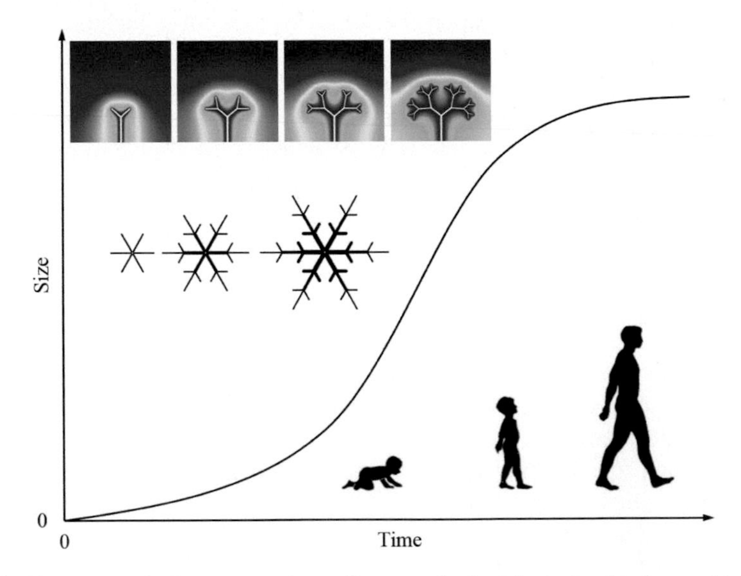

Fig. 8.5 Every growth phenomenon (spreading or collecting) is destined to have a history represented by an S-curve when plotted as size versus time: animal bodies, snowflakes, and trees everywhere

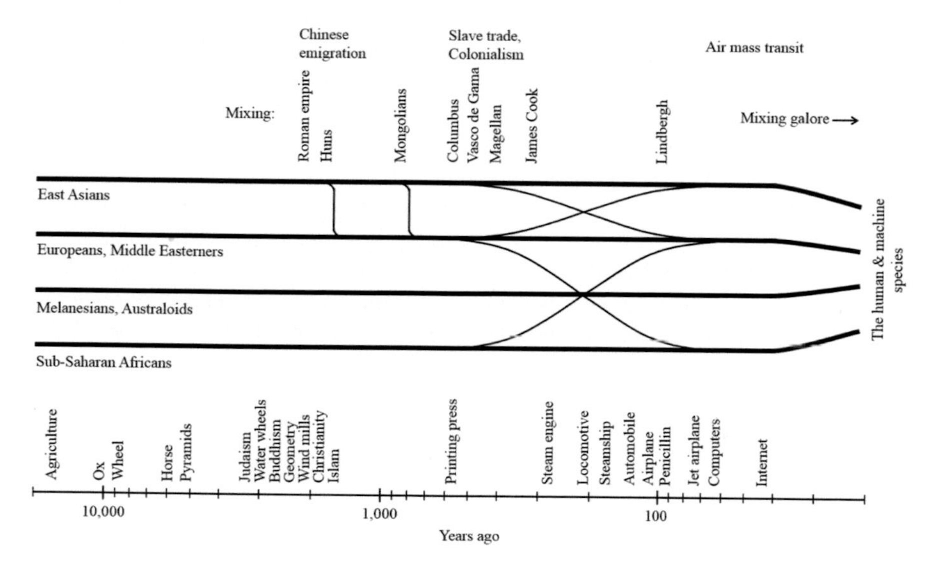

Fig. 8.6 The evolution toward greater movement and human mixing on the globe: from the modern humans that concluded Fig. 8.3, to the human and machine species of today

The most glaring (and most overlooked) difference between the two phenomena is that the "evolution" sketched on the horizontal is discontinuous, while on the vertical every single "growth" is a continuous movie of growing, morphing, and shrinking images, each from one birth to one death. On the horizontal, we see the familiar alignment of animal shapes, which unfortunately (or intentionally) has been creating the illusion that somehow, like a "transformer" toy, the ape got up on two legs and while walking it morphed into a modern human.

No, and the finite timeline of the Neanderthals (Fig. 8.3) makes it very clear: every adult that populated discretely the horizontal line in Fig. 8.4 became extinct long ago, and that includes the solitary, naked, cold, and hungry *Homo sapiens*. The modern man that spread all over the globe is the *human and machine* species that today drives off the page in an automobile, to the right, out of Africa too.

In the name of our species, the word "machine" carries its original meaning: a sophisticated contrivance that allows for more effective use of human effort (from the word *mihaní* in old Greek). Without our add-ons, we are so weak that we would all vanish in less than one lifetime. According to Frank Zappa, "If your children ever find out how lame you really are, they'll murder you in your sleep."

Growth is continuous because it is a spreading flow (or a collecting flow) with a future that is S-shaped [7, 8] when plotted as size versus time (Fig. 8.5). This is illustrated as the growth of one animal, one snowflake, and one tree-shaped flow architecture in the garden, river delta, and spilled milk. The S-curve past and future of any growth is true across the board, in all the animate, inanimate, social, and engineered flow architectures that have freedom to morph and space to spread.

During the growth of the human population on the globe, the spreading flow mixed with what it encountered as it flowed over the area. This mixing accounts for the very slight differences in body and face architecture, from Africa to the Middle East, Europe, Melanesia, and East Asia.

"The grass is always greener on the other side" is the constructal-law urge that drove and continues to drive human migration on the globe. During the Dark Ages, the Asiatic groups (from the Huns, Bulgars, and Hungarians, to the Mongols, Tatars, and Turks) migrated westward because of the greener pastures. Why, because at the same latitude the west is greener because the warmer and more humid winds blow from the Atlantic in the easterly direction. At the same time, and for the same reason, the European shepherds and farmers did not migrate to the east. Migration happens one way, toward easier, freer, and longer life. Today, we see this clearly in how people migrate toward more freedom. One way, or no way.

I do not know what the UK student did with my advice, I know only that I look my own advice seriously. This is how I drew Fig. 8.3, which is based on reading Refs. [9–15]. Europeans, Middle Easterners, and East Asians are not pure *Homo sapiens*. They are mutts. The pure specimens of *Homo sapiens* still populate the continent south of the Sahara.

The DNA of other hominids—Neanderthals and Denisovans—has not vanished entirely from the surface of the earth. It is in every body that originated from east and north of the Sahara. The percent of Neanderthal DNA in non-Africans varies between 1 and 4% [12]. In this narrow range, East Asians have generally 15–30% more Neanderthal ancestry than Europeans [13].

The compounding of interbreeding events left its mark on several features of human body morphology, all aligned in time with the succession of interbreeding events, from none in Sub-Saharan Africans to one interbreeding event in Melanesians, two events in Europeans and Middle Easterners, and three events in East Asians. With each interbreeding event, the center of gravity of the body shifted lower, the limbs became shorter, the body height decreased, the face became rounder, the legs occupied a smaller fraction of the body volume (more on this in Fig. 9.4), the brain cavity increased in relation to body size, and the cross section of the shaft of human hair diversified from flat in Africans, to oval in Europeans and Middle Easterners, and finally round in East Asians (or, in the same time sequence, from kinky hair to wavy hair and finally straight hair).

All together, the modern humans that exited from Fig. 8.3 to the right continued to spread and mix on the globe, as we see next in Fig. 8.6. The spreading and mixing were aided immensely by the spreading of boating, agriculture, domestication, language, science, and technology, which made long-distance travel accessible to larger and larger groups. What we see today is just the fastest phase (the middle of the S-curve) of this spreading flow, which for the past two hundred years has been riding on the train of power from fire and the industrial revolution.

More diversity ensued because en route the train discovered its own science, the science of engineering with its expanding number of disciplines. Figure 8.7 shows this body of science from its origins in antiquity to the muscular activity that propels the world today. Mechanical engineering was the new science of "machines,"

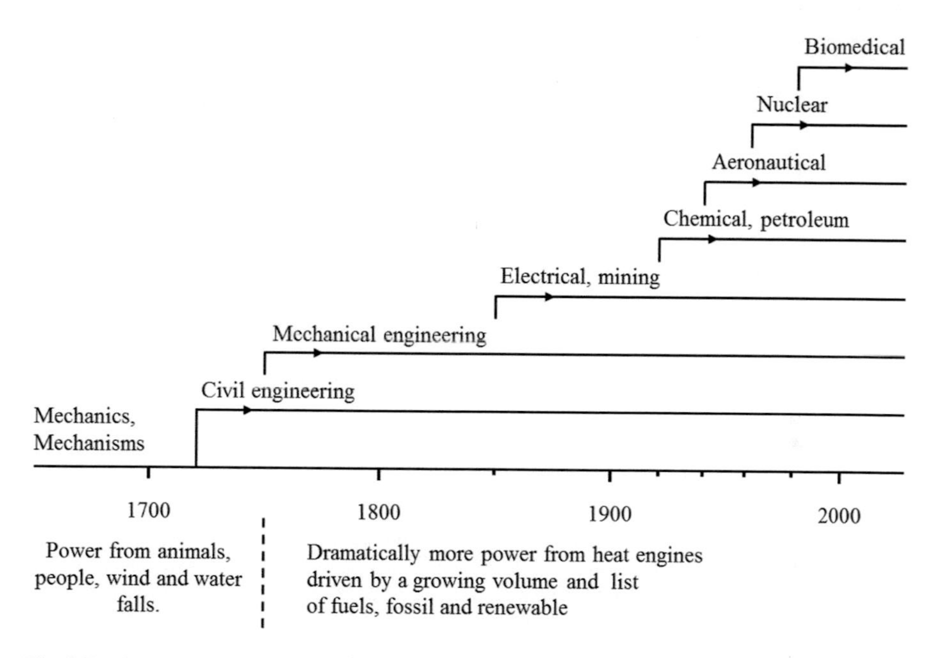

Fig. 8.7 The evolution, spreading and specialization of engineering disciplines during the past two centuries

the machines driven by heating from burning fuels. What was before mechanical engineering acquired the name of "civil" engineering (the constructing of living in the city), even though most of that science was invented in antiquity and the middle ages for military purposes: the mechanics of roads, bridges, ramparts, catapults, weapons, and military campaigns. Engineering education was first military education, as we are reminded today by the École Polytechnique in Paris, located in the Latin Quarter originally, which was the first engineering university in the world.

Newer kinds of engineering emerged as natural add-ons because of dramatic changes in the technology of more power for more people. Electrical engineering happened because the power generated from burning fuels was needed by greater numbers of users on areas (city, countryside) situated far from the steam engine driven by fire and the turbine driven by the waterfall. The mechanical power was converted into electrical power, which was then spread on the area and used directly (for lighting and heating) and converted back to mechanical power, for transportation and manufacturing.

Chemical and petroleum engineering became distinct disciplines at the turn of the previous century, as the demand for new fuels in bigger quantities became dominant. Aeronautical engineering took flight during the First World War because of the military importance of human flight. Nuclear engineering was born out of military need during the Second World War. Biomedical engineering is now riding high in university education and modern hospitals, primarily because affluence (wealth) and many new technologies have made it easier to improve the human body by design.

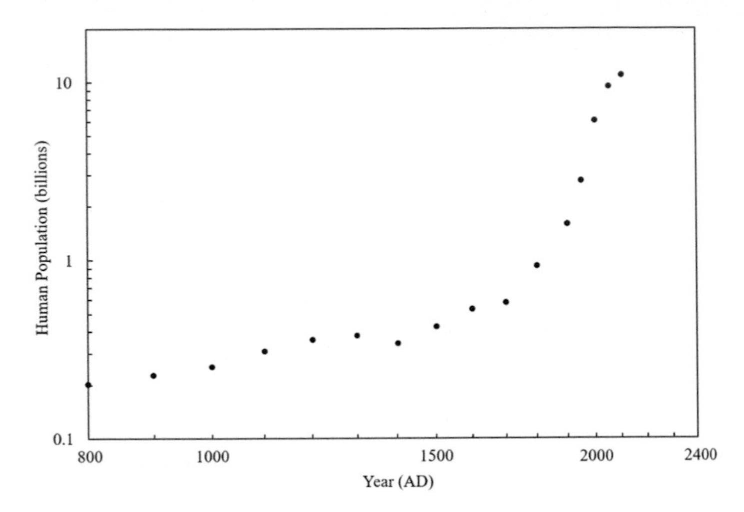

Fig. 8.8 The S-curve growth of human population on the globe, or the "Cambrian explosion" of new human and machine species triggered by the sudden access to power from fuels. The data were read from the video available in Ref. [16]

Still, biomedical engineering is as old as civil engineering, and its roots too are military: shields, helmets, bandages, and prostheses.

Power and the expansion of know-how are making all of us more advanced, more free, wealthier, and longer living. Power from engineering is why the human population has grown so sharply on the globe [16] (Fig. 8.8). With power, we have been spreading not as the naked humans of Figs. 8.3, 8.4, 8.5 and 8.6, but as the fast improving and diversifying human and machine species [17, 18]. All the spreading flow architectures, however, are commanded by physics to have an S-shaped history of how their territory grows over time, cf., Fig. 8.5. The S-curve plateau of the future of the human population is already evident in Fig. 8.8.

The defining effect that machine power (the engine) has on the spreading of humans is analogous to the effect that the organ for vision (the eye) had on the spreading of animals on the globe. The emergence of vision triggered the Cambrian explosion of new species that invaded the earth 541 million years ago. With vision, the animal could explore its surroundings (for food, safety, mate, and shelter) far deeper, and with much greater safety. The emergence of the eye as a sensory organ was far more empowering than the touch. The evolutionary change in movement on earth occurred in the constructal-law direction, from the biosphere without vision to the biosphere with vision.

The industrial revolution, with Internet and AI, is the new Cambrian explosion. The original Cambrian explosion toward more vision continues today in the evolution of technologies for transportation, surveillance, and warfare. Vision, or the knowledge that vision makes possible, is the ability to see ahead in space and in time. Try to imagine ancient and medieval defenses and warfare without fires on earth mounds to warn of danger. Imagine submarine navigation without periscope and sonar. Think of aviation without radar in the 1940s, and without GPS today. The living thing, animal or human, lives longer, travels farther, and faster with a better organ for vision. This evolutionary direction is the time arrow of the constructal law, and it is why the newer (the birds) have better eyes than the older (the fish), and why predators in all media have better vision and greater speed than their prey.

Power today is the new Cambrian explosion, in fact, this explosion dwarfs the effect that the Cambrian explosion had on changing the earth's crust. The human discovery of how to generate power from inanimate surroundings such as fire [19] 300 years ago has been liberating for everyone, not only for slaves and work animals. The constructal-law urge in each of us is to move more, more easily and farther into the available space. In physics, all this means more power, i.e., the urge to have access to more power to move things (Fig. 1.5).

Here is how the "power Cambrian explosion" is generating new human and machine specimens, with astonishing numbers and diversity. All it takes is a look at the time chart in Fig. 8.7. In short, this chart is the evolution of *engineering*, which according to the dictionary is a *science*: the science of how to put physics principles in the service of humans. The contrivances that empower us today have many uses, ages, configurations, sizes, and numbers. They are diverse, and at first sight they appear complicated, disorganized. That is not the case. The contrivances (our machines, literally) come from a continuous phenomenon of free evolution of the human and machine species toward more power and, as a consequence, more life.

Early in our civilization, the power came from domesticated animals and humans equipped with wheels, carts, and wheelbarrows. Road building, catapults, canals, and water wheels enhanced human movement, from antiquity into the middle ages. In retrospect, we call this useful science civil engineering, although its name and place in science is much bigger: it is *mechanics*, the physics of the contrivances that carry forces, move, support, and carry us.

People achieved great things with mechanics, and achievements continue today. One example is the evolution of handwriting, from the stylus on slate to the pencil, the fountain pen, and now the throw-away bic pen. Another is the evolution of construction, bridges, roads, and buildings. The designs, their shapes and sizes, evolved with their materials and knowledge (methods), from wood, mud, and dry stone construction, to fired brick, concrete, steel, and glass.

The power used for construction underwent an evolution of its own, and here is where the Cambrian explosion happened again. When people's access to more power (more watts, the ability to move something to a new position on earth, against forces that oppose the movement) became much greater because of the motive power of fire, there was an explosion in number of new types of contrivances that empower the modern humans. New classes of contrivances became possible, so many, so efficient and useful, and so diverse that we take their mother (the engine) for granted. Most of us who have access to electric power today have no idea how the power got there, in the electric outlet. Only earthquakes, hurricanes, and despotism (lack of freedom) would wake people up to the liberating role that engineering plays in our lives today.

The power explosion is documented for posterity not only in the discarded objects taken to garbage dumps, rivers, and oceans, but also in the changes in how we teach the science of useful ideas and processes. When power and transportation from burning coal and wood became available from machines, the old mechanics grew the new branch called mechanical engineering. The old contrivances, roads, and bridges continued to be made with power from fire, not with power from animals and slaves: this became civil engineering. It was only after the fire-driven engine and mechanical engineering that civil engineering emerged as the oldest, the primordial useful science.

One hundred years after James Watt, the new class of contrivances and their science became electrical engineering. This extension came naturally, as a consequence of the success of power generation through mechanical engineering. More power was made available for more people, and more people live farther from the power source. The challenge was to distribute the swelling stream of power on the landscape, to empower larger and larger groups. Enter Tesla and Westinghouse, and you see how electrical engineering came into being. It is the science of converting mechanical power into electrical power so that it can be distributed to great distances where it can be converted back to mechanical power and used (destroyed) in the course of moving earth and vehicles, and changing the world.

With power generation, the conversion and distribution of anything is now possible, including the easy access to power. Once precious in the functioning of the first locomotive, light bulb, and electric heater, power is now derisory. On the back of the humble flow of power into our homes, schools, factories, and hospitals are growing forests of new and highly diverse contrivances, each belonging to a new class and a new science. After mechanical and electrical engineering, the new contrivances added new layers to engineering science—the science of the useful—and to the organization of the university: review the timeline in Fig. 8.7.

In all the compartments of knowledge, people are contriving to make each of us a more powerful and longer living human and machine specimen. This is no exaggeration. The old man with hip implants and hearing aids is young compared with the deaf man limping in pain. Think of this eventuality and recognize your great fortune as a carrier of the contrivances and knowledge that came to you from the science of useful things. They came to you not because you are deserving, but simply because you were lucky to be born in the advanced society built by others, humble and modest, and hungry too.

We cannot have enough of the "machine" in the *Homo*. To fear that the machine will overpower, the *Homo* is strange. On the contrary, the machine liberates and empowers the *Homo*. Overpowering happens daily: the new human and machine specimen overpowers the current human and machine specimen. That's evolution, the evolution of contemporary and future humans.

The human and machine species is just the latest, the more prevalent and omnipotent manifestation of the broad phenomenon of biological evolution called "niche construction." Like the wall cavity in which the sculptor protects the sculpture (from *niche* in French), many animals shape and organize their immediate surroundings to facilitate their living. The spider web, the bird nest, the nuts buried by the squirrel, and the beaver dam are of the same nature and built with the same purpose as the dam for the hydroelectric power plant.

In view of all this, biology is not about animal evolution here, and niche construction evolution there. It is about the evolution of an immense diversity of *animal and niche* species. Likewise, anthropology is not about modern human evolution (Figs. 8.3, 8.4, 8.5 and 8.6) and, separately, technology evolution (Figs. 8.6 and 8.7). It is about the evolution of one species, the *human and machine* species.

I end with an example of how the story of this chapter empowers us to answer "tough questions." The producer of a BBC comedy show called me recently with a tough question, which they were planning to pose to a panel of celebrities on stage: If civilization were to have evolved without one of its two greatest inventions, the wheel and the printing press, which of the two would have made us more advanced today? In other words, which is more essential?

The answer is easy coming from the constructal law, but the question on the telephone was so unexpected and the caller so funny that I thought immediately of two instances in which images of the wheel and the printing press made me laugh.

First, the wheel. There are artists who remind us with drawings and even photographs that in antiquity the wheel was solid and heavy looking. So, what do these artists do? They show us the image of a millstone. An old millstone is easier to find, but one cannot even lift it because it is heavy. The reader is put to the test by the ignorance of the artist.

Second, the press. A university library was celebrating 500 years of the printing press and the Gutenberg Bible with facsimiles of pages of the old book, including an old press put on display right at the entrance to the library. All was good looking, except that the old press was a wine press, a cylindrical basket with vertical wooden bars for squeezing grapes and squirting juice, not for printing books.

The answer to the comedy show question is the wheel, not the printing press, because the history of our civilization is our evolution toward greater and easier movement of human bodies and artifacts on the globe. Two designs dominate the evolution of our movement: the evolution toward more power generation and use (from work animals to steam engines and many more kinds of power), and toward better (more economical, efficient) and lighter vehicles (from sleds and carts to automobiles and airplanes).

The wheel represented a dramatic jump in the evolution toward greater movement for humanity on earth. Relative to the sled, the improvement in ease of movement on the landscape was epochal. The only other step that competes with the wheel in importance in the evolution of human movement was the steam engine and the industrial revolution, yet, to have an engine one needed to have the wheel first. Together, the wheel and the engine led to freedom, empowerment, safety, advancement, culture and wealth, and this way to even more freedom.

The printing press was just one in a voluminous diversity of artifacts that make us more powerful and efficient (lighter, faster, safer, etc.), as we tend to move and communicate on the globe. Other contrivances from before and after the printing press were fire, alphabet, money, weapons, science, telegraph, radio, television, internet, etc. All technology evolution is about this.

For the comedy show, the harder question would have been to choose between the printing press and the wine press.

References

1. A. Bejan, *Convection Heat Transfer*, 4th edn. (Wiley, Hoboken, 2013)
2. Why every snowflake is NOT unique, Duke University, December 20, 2013. https://www.youube.com/watch?v=8hFo23ZJ6YU
3. M. Zhilin, S. Savchenko, S. Hansen, K.-U. Heussner, T. Terberger, Early art in the Urals: new research on the wooden sculpture from Shigir. Antiquity **92**(362), 334–350 (2018)
4. A. Bejan, J.H. Marden, Unifying constructal theory for scale effects in running, swimming and flying. J. Exp. Biol. **209**, 238–248 (2006)
5. J.D. Charles, A. Bejan, The evolution of speed, size and shape in modern athletics. J. Exp. Biol. **212**, 2419–2425 (2009)
6. A. Bejan, E.C. Jones, J.D. Charles, The evolution of speed in athletics: why the fastest runners are black and the swimmers white. Int. J. Design Nature Ecodyn. **5**(3), 199–211 (2010)
7. A. Bejan, S. Lorente, The constructal law origin of the logistics S curve. J. Appl. Phys. **110**, 024901 (2011)
8. A. Bejan, S. Lorente, The physis of spreading ideas. Int. J. Heat Mass Transf. **55**, 802–807 (2012)
9. K.L. Beals, C.L. Smith, S.M. Dodd, Brain size, cranial morphology, climate, and time machines. Curr. Anthropol. **25**(3), 301–330 (1984)
10. C. Stringer, Modern human origins: progress and prospects. Philos. Trans. Royal Soc. B **357**(1420), 563–579 (2002)
11. C. Stringer, The origin and evolution of *Homo sapiens*. Philos. Trans. Royal Soc. B **371**, 20150237 (2016)
12. R.E. Green et al., A draft sequence of the Neanderthal genome. Science 328, 710–722 (2010)
13. B.Y. Kim, K.E. Lohmueller, Selection and reduced population size cannot explain higher amounts of Neanderthal Ancestry in East Asian than in European human populations. Am. J. Human Genet. **96**, 454–461 (2015)
14. G. Roth, U. Dicke, Evolution of the brain and intelligence. TRENDS Cognit. Sci. **9**(5), 250–257 (2005)
15. R.D. Martin, Relative brain size and basal metabolic rate in terrestrial vertebrates. Nature 293, 57–60, 3 September 1981
16. How we became more than 7 billion—humanity's population explosion, visualized, American Museum of Natural History, Aeon, 2 December 2016
17. A. Bejan, *Shape and structure, from engineering to nature* (Cambridge University Press, Cambridge, UK, 2000)
18. A. Bejan, *The physics of life: the evolution of everything* (St. Martin's Press, New York, 2016)
19. A. Bejan, Why humans build fires shaped the same way. Nature Sci Rep **5**, 11270 (2015)

Chapter 9
Evolution

With freedom comes evolution, and with evolution come all these visible things, complexity, diversity, hierarchy, size, and seemingly free choices that speak of the universality of economies of scale. People often ask me to predict something that will happen. Can evolution be predicted?

Evolution can be predicted by invoking the law of physics that governs the evolution phenomenon. It is the same with other phenomena that are universal. For example, by invoking Newton's second law of motion, we can predict movement in all its forms, past and future, from the movement of planets to the trajectories of projectiles of all sizes (bullets, water droplets) and the continuous flows of fluid everywhere. With the second law of thermodynamics (the law of irreversibility), we can predict that all movement (inanimate, animate, machine) will be dissipative (imperfect), and oriented one way if flowing by itself. By calculating the irreversibility (quantitatively, as the entropy generation rate), we can measure the difference between the real system with currents that flow one way, and the ideal system with currents that theoretically (in the limit) could flow unchanged and by themselves in both directions.

Even though it sounds like a contradiction, with a law of physics we can predict the past of a phenomenon. Why, because before the law was known there was no reason to investigate the past in order to test the law. The predictive power of the law has been demonstrated multiple times over the past two decades, which is why the law is now well established. The predicting goes this way:

First, by invoking the constructal law, the thinker's mind "sees" an evolving (morphing) flow architecture—its physical characteristics and direction of changes over time—and anticipates the main features (for example, the scaling trends and even the formulas) of the architecture that "will be." This first step is theory, or more precisely, the constructal theory of the particular evolutionary design that is being contemplated. One theory for one phenomenon.

The word "theory" must be used only when it is applicable. According to the dictionary, theory is a purely mental viewing, contemplation, or speculative idea of

A. Bejan, *Freedom and Evolution*,
https://doi.org/10.1007/978-3-030-34009-4_9

how something might be. Theory is the mental flash in the dark of the night. It is not a piece of mathematical analysis. Theory is not the practice of looking at nature, recording observations, and reporting measurements. That practice is empiricism, also known as monitoring, modeling, experimenting, tinkering, correlating, mimicking, and copying (as in biomimetics, p. 117). With theory, the human mind first sees the mental image—the new mental connection, the innovation (cf. Chap. 5)—and only later it checks the validity of the prediction by comparing it with observations of nature.

Theory is mental viewing first, and comparison with observations of nature later. The time arrow of empiricism is the opposite: observations first, which are followed in time by ideas about condensing, storing, and reporting observations. Theory must not be confused with empiricism. They are opposites. There is no such thing as "theoretical model."

Second, after the theory other thinkers compare the predictions with the measurements that are available. If observations are not available, then new investigators will devise new experiments that will generate new data in order to check the validity of the predictions. After all, this is the purpose of experiment; it is to verify the correctness of an idea, a theory, a purely mental viewing. Experiment without an idea to define and justify the experiment is nonsense or, at best, a make-work project.

Many researchers have compared constructal theories with existing data, and with new data that came from their laboratories and library searches, after the theory. Not surprisingly, constructal theory papers begin with the predictions (the theory) and continue with the clouds of data that fall right on top of the theoretically predicted curves. This, by the way, is how to discover the correct curves that correlate the data, the allometric rules (cf. Chap. 7). This technique is general and useful. I proposed it in 1984, in the first edition of my book *Convection Heat Transfer*: if you want to discover the correct correlation and the correct dimensionless groups that summarize a phenomenon, first you must construct the theory that predicts the correlation and its proper dimensionless groups. This boils down to the need to understand the physics before you embark on measuring and correlating data.

Following the publication of the constructal law of 1996, the examples of its predictive power have been multiplying. Constructal law was used to predict many phenomena that seem unrelated: for a broad review with peer comments, see Refs. [1–10]. Here are just three examples taken from my publications, one animate, another inanimate, and the third from the human realm (technology evolution).

Animal locomotion. First, we predicted the formulas for the speeds, frequencies, and power requirements of all animal movers in all media (water, air, land). Later, by comparison with a very large volume of zoology data, the formulas became the physics-grounded correlations that underpin animal design. This theory was first proposed and validated (for fliers only) in my 2000 book [11], after which it was extended to swimmers and runners [12], Fig. 9.1.

In brief, easier access for the body that moves through its environment (water, land, air) means that at every time step a balance must be reached between the work spent on the vertical (to lift the weight that keeps falling under gravity) and the work

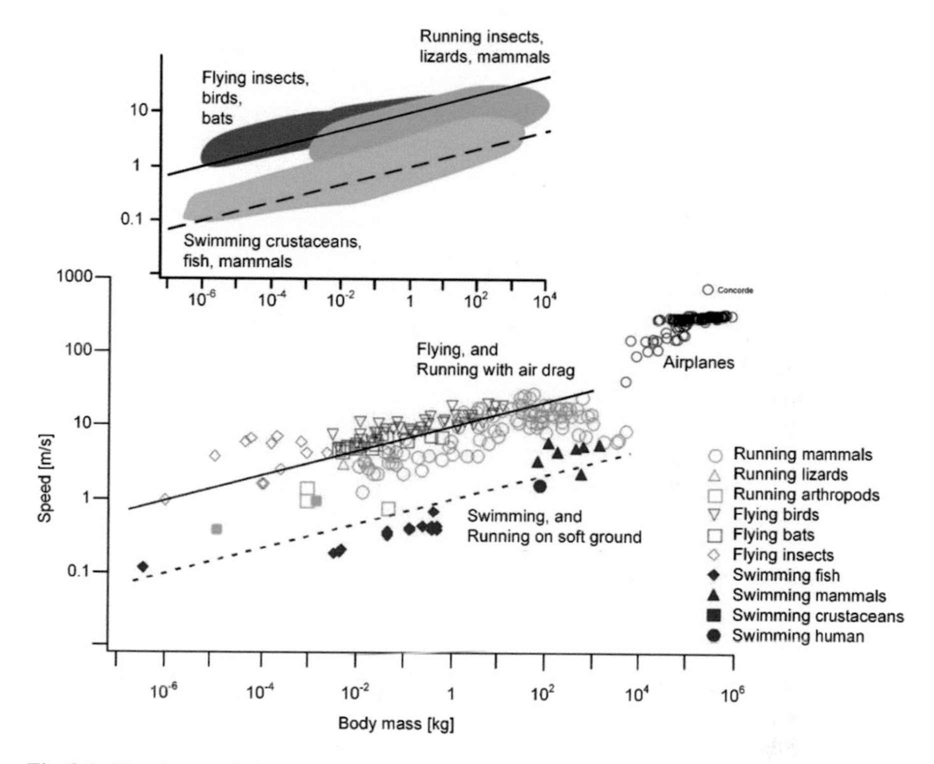

Fig. 9.1 The characteristic speeds of all the bodies that fly, run, and swim. The upper frame offers a bird's-eye view of the animal data put on display in the lower graph

spent on the horizontal (to get the environment out of the way). From this balance follow theoretically the most basic and best known facts about animal locomotion: bigger animals should be faster, should undulate their bodies less frequently, and should be stronger. This includes the swimmers, which must lift water in order to advance forward. The water displaced by the swimmer can only go up because the upper surface of the body of water is freely deformable while the bottom and the sides are rigid.

Evolving in accordance with the same principle is the locomotion of the human and machine species. The evolution of athletics [13–15] is one laboratory where the constructal law of animal locomotion has been tested successfully based on the recorded evolution of winning speeds in running and swimming, short distance and long distance, and men and women.

All locomotion is a rhythm, because all creatures on Earth locomote in the presence of gravity. The rhythm is the one–two that comes from balancing better and better the time spent on the vertical (to lift the body) and the time spent on the horizontal (to push the body forward). Freedom makes such choices possible.

To feel the need for such rhythm, imagine that you are crossing the Danube in a rowboat that is leaky. You must divide your time between pulling on the oars and scooping water from the bottom of the boat and dumping it overboard. If you don't scoop, you sink and never reach the other shore. If you scoop all the time and don't pull the oars, the boat stays dry but you never leave the shore.

River basins. First was the law-based prediction that any river basin on the globe should have on an average a modular structure with approximately four tributaries to a mother channel. Next, the prediction was compared with data correlated empirically by Horton and others since the 1930s [16–18], Table 9.1. The prediction and the validation were published together in 2006 [19], on the following theoretical basis:

The constructal-law field began in June 1996 with imagining "the city" and formulating the fundamental "area-point access" problem in terms of people flowing through streets of various sizes [20–22]. Chief is the "modularity" of the construction of area-point flow architecture for access, which was demonstrated for heat conduction cooling, urban design (streets), and fluid flow through ducts and porous media.

The river basin construction is the constructal theory of the area-point design of a natural area-point flow system that everybody knows. The objective of this particular theory was to uncover the rule of construction that offers the entire basin access from area to point, regardless of its size (in Table 9.1 the size of the basin is indicated in the i column). Four rules of construction were evaluated sequentially, from left to right in Fig. 9.2:

(a) River basin with four area elements A_0. To build the area $4A_0$, one has to execute two steps of doubling, or two pairing events, from A_0 to $2A_0$, and then from $2A_0$ to $4A_0$. In this construction, there is only one rule, doubling, or "the rule of 2."
(b) River basin with four area elements A_0. To build the area $4A_0$, one has to assemble four A_0 elements in a single step. This step is one of quadrupling, or "the rule of 4."
(c) Eight A_0 elements are assembled in two steps: quadrupling, which gives birth to $4A_0$ constructs, and then doubling, which joins two $4A_0$ constructs into one $8A_0$ construct.
(d) Eight A_0 elements are built into one $8A_0$ construct in one step. This is "the rule of 8."

By constructing and comparing (a)–(d), we found that the construction rule (b) offered the lower global pressure drop, which for rivers means a smaller altitude difference between inlets and outlet on the area, or easier flowing and greater access to what flows.

The construction uncovered in Fig. 9.2 is the origin of all *vascular* designs, inanimate and animate, in two and three dimensions [23]. The constructal-law phenomenon is why vascularization happens in so many flow systems, from river basins to the nervous system and the brain.

Table 9.1 The theoretical architecture of constructal river basins

i	A_i/A_0	$L_{Ti}/A_0^{1/2}$	N_i	R_{Li}	R_{Bi}	$D_{\omega i}A_0^{1/2}$	$F_{si}A_0$	$F_{si}/D_{\omega i}^2$	$L_{Mi}/A_i^{1/2}$
0	1	$\frac{1}{2}$	1	–	–	$\frac{1}{2}$	1	4	$\frac{1}{2}$
1	4	$\frac{7}{2}$	5	3	4	$\frac{7}{8}$	$\frac{5}{4}$	1.63	$\frac{3}{4}$
2	4^2	$\frac{35}{2}$	21	2	4	$\frac{35}{22}$	$\frac{21}{16}$	1.10	$\frac{3}{4}$
3	4^3	76	85	2	4	$\frac{76}{64}$	$\frac{85}{64}$	0.94	$\frac{3}{4}$
4	4^4	316	341	2	4	$\frac{316}{256}$	$\frac{341}{256}$	0.87	$\frac{3}{4}$
River basins				1.5–3.5 Horton	3–5 Horton			0.7 Melton	~1.4 Hack

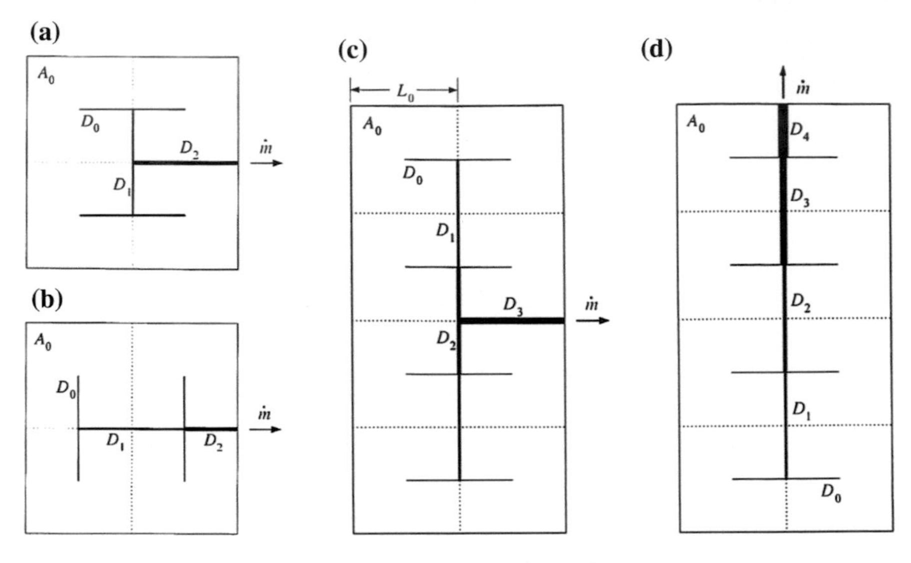

Fig. 9.2 Four constructions of river basins consisting of area elements

The example of river basins (Table 9.1, Fig. 9.2) teaches the most fundamental characteristic of any work that is theoretical. With a law of physics, one does not "explain." One *predicts* by invoking the law, without looking at an actual object in nature. For example, in constructing Fig. 9.2, we invoked the constructal law in order to identify (b), and in this way we predicted that the counting of geometric features in all river basins in geophysics publishing history should turn out the way that our (b) construction recommends. This was validated by comparison with Refs. [16–18], as shown at the bottom of Table 9.1.

We do not know why the "whole" area has the natural tendency to evolve into a hierarchical flow construction that offers access greater than other constructions. All we know is that this natural tendency (along with other evolutionary designs, lungs, snowflakes, and city streets) is predictable if one invokes the law of physics of evolutionary design.

River erosion is akin to self-lubrication, as in the lava flowing through volcano shafts that acquire round cross sections [24], and in rolling stones that become rounder as they roll [25]. All these phenomena are manifestations of the constructal law and are predictable by invoking the law. In river basin evolution "erosion" underpins all the morphing that goes on:

(i) In every channel, the river cross section channel becomes enlarged over time.
(ii) Every channel cross section becomes lemon slice shaped, regardless of channel size, so that all river channels that are wide are also deep. The law commands that the ratio width/depth should be a universal proportionality for river channels of all sizes [11].

(iii) The whole exhibits the natural tendency to organize and put its channels and area elements into seemingly modular constructions that are there for global flow access from area to point.

There is no "model" in the thinking that generated Fig. 9.2 and Table 9.1. We were not looking at a river map when we made the four drawings shown in the figure. The different channel thicknesses visible as thin and thick lines were derived from the constructal law, not from observations of nature. For the theorist, the page was initially blank, and on the page the theorist morphed the flow architecture while not knowing beforehand which architecture would be "naturally selected" by the constructal law.

By identifying the essential features of the flow architectures endowed with global access, we predict the direction of evolution over time. We do not predict an end design, because end design does not exist. With the law of physics of evolution, we predict the direction of evolutionary changes over time.

Aircraft. First, we predicted the formulas that express relations between dimension of organs and body size in small and large airplanes. After the theory, we compared the predictions with the voluminous body of data accumulated during the entire history of commercial aviation [26]. When plotted as recommended by the theory, the data aligned themselves. Every time, the agreement between the prediction and the alignment of its respective data was consistently excellent, a surprise for the reader maybe, but not for the theorist.

More recently, we extended this theory to predicting the evolutionary future of helicopters [27]. The same theory also allowed us to provide a physics basis for the surprising alignment of performance data from the early days of aviation [28], Fig. 9.3. Airplane technology is an example of how a newly formulated law of physics predicts the past of an evolutionary phenomenon by unifying the old volume of data with the new data.

In summary, there is one phenomenon (evolution), and one law of evolution. The theories that spring from the law are as many as the circumstances in which the thinker contemplates one phenomenon, which is just one natural manifestation of the law. There is one constructal law, and there are many constructal theories that cover the board, from biological to non-biological phenomena, and from the tiny pulmonary alveolus to big celestial bodies, lung structure, rhythm (respiration, heartbeat), animal locomotion, river basin structure, river channel cross sections, aircraft evolution, turbulent structure evolution, snowflake evolution, and many more. Theory is not Law.

While discussing the validation of the theory that celestial bodies should exhibit hierarchy—few large and many small bodies suspended together and attracting each other [29], instead of one size—we addressed Ellis and Silk's test [30] that "the issue boils down to clarifying one question: What potential observational or experimental evidence is there that would persuade you that the theory is wrong and lead you to abandoning it? If there is none, it is not a scientific theory." There are numerous tests of evolutionary design (in addition to the three discussed already, Figs. 9.1, 9.2 and

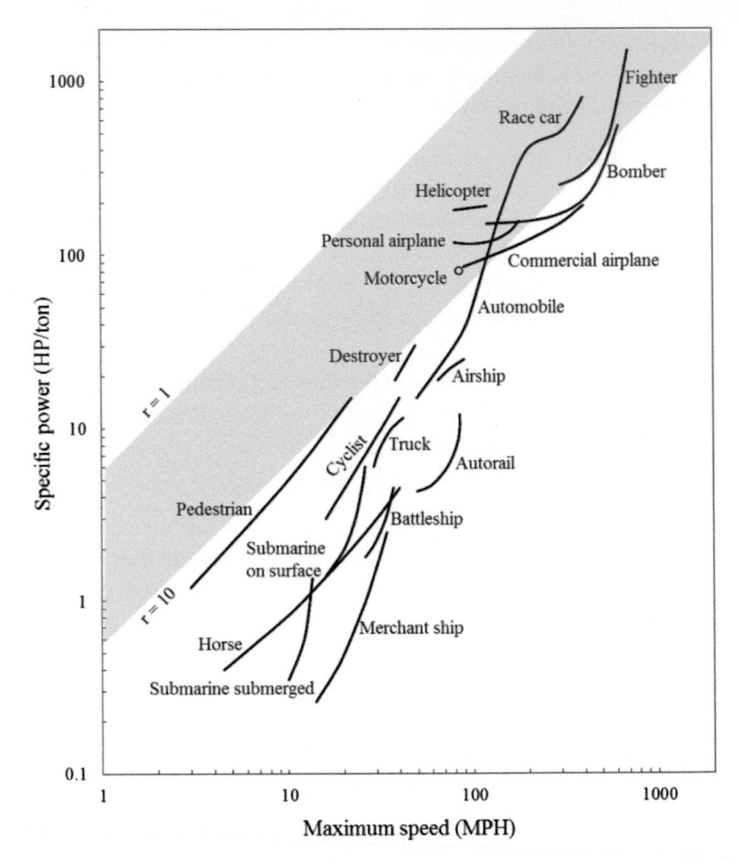

Fig. 9.3 Specific power of single vehicle from before 1950. The dark band represents the domain predicted by the constructal theory of all locomotion, with $r = 1$ for air, $r = 10$ for water, and land in-between

9.3) that validate the constructal law, in addition of the hierarchy of celestial bodies, for example:

- A flat plume or jet should evolve into a round plume or jet, never the other way around [31].
- Solid bodies that grow during rapid solidification (e.g., snowflakes, dust clusters) should be tree-like, not spherical [32, 33].
- Bigger moving bodies (animals, rivers, vehicles) should live longer and travel farther, not live less and travel less [34].
- The human lung should be tree-shaped with 23 levels of branching [35]. The number 23 emerges predictively from the trade-off between the time spent by air as it flows through hierarchical air tubes, and the time spent by the diffusion of O_2 (and CO_2) across the blood vascularized tissues that form the walls of the alveolar sacks. The trade-off is such that the sum of the two times (tube flow + diffusion)

is minimal when the number of branching levels is 23. In an animal smaller than the human, the number of branching levels is predictably smaller.

- All animal speeds (swimming, running, flying) should be proportional to the body size raised to the power 1/6, and, for a given body size, should increase from swimmers to runners and then flyers [12].

The examples go on, and every single one is evidence in answer to the question formulated by Ellis and Silk. There is more to this than meets the eye. Some readers may be tempted to argue that the evidence has long been available in the past, and that the theory cannot predict past observations. This argument is incorrect, in two ways:

First, predicting old observations that were not recognized and questioned previously is indeed theory. Examples are Galilei's law of gravitational fall, and Clausius' law of irreversibility (the second law). The fact that everything on earth has weight (*gravitas* in Latin, from which the word "gravity") and the fact that everything flows by itself from high to low (one way, from which "irreversibility") were old and familiar phenomena that were not brought into science before Galilei and Clausius questioned them and summarized them in the form of two succinct statements.

The obvious fact that animals of the same kingdom look different was not questioned before I predicted that in bigger terrestrial animals the legs (i.e., the lifting organs) should occupy a larger fraction of the body volume than in smaller terrestrials (cf. Fig. 5.2 in Ref. [36]). Now, in Fig. 9.4, I show that the same should be true in all animals, fliers, runners, and swimmers. The lifting organs of the flier are the wings, and the lifting organ of the swimmer is the undulating body. I drew all these animals so that on the page they have the same size, even though the animals in the right column are enormous in comparison with the animals in the left column.

Every size has its own design. Said another way, design is another name for size. The *diversity* that is evident in Fig. 9.4 is a manifestation of the same physics principle as the human diversity detailed in Chap. 8. Accordingly, taller humans should distinguish themselves as having bodies in which the legs constitute a bigger fraction of the body.

Second, throughout the physics literature there are numerous predictions that refer to *future* observations, such as the examples given above for plumes [31], rolling stones [25], and snowflakes [32], and many more evolutionary designs that occur at short time scales that are comparable with the duration of one human life, for example, the evolution of technology [26–28] and athletics [13–15]. This should come as no surprise because all science is an artifact (an add-on to the mind, the brain design) that empowers humans to predict the future for human benefit.

Even more support for the predictive power of evolution as physics is available under different names. Biomimetics, or biomimicry, is portrayed as a new science the core idea of which is that the most efficient systems are found in life (bio) forms. First, "most efficient" cannot be true because the superlative "most" implies that life forms have stopped evolving. Second, "efficient" does not belong with imitating (mimicry). The concept of "efficient" requires an understanding, which happens in a mind endowed with knowledge, which is ideas, principles, and what to do with ideas

Fig. 9.4 In bigger animals in all media a larger fraction of the body mass is occupied by the lifting organs, wings in the air, legs on land, and undulating body in water (Drawing by Adrian Bejan)

and principles. Fashionable is also the claim that biomimetrics helps technology. Biomimetics does not, but the principle does. Biomimetics is successful only when the observer knows the principle that accounts for the observed natural object. Those who claim success with biomimetics are relying intuitively on the physics principle of evolution and design in nature. I was writing these observations while watching an elephant on the Lower Zambezi, as it was shaking the ana tree so that the dried pods would fall on the ground. The elephant discovered this knowledge on its own, not by mimicking the blowing wind that also shakes the tree. The elephant does not blow and does not mimic humans by throwing sticks at branches with fruit. Likewise, humans discovered that round pipes are good for water flow. They did not copy the cross sections of blood veins and pulmonary air tubes in order to make pipes.

The elephant does not rise on its hind legs to reach higher. The giraffe does not do that either. Getting up on two legs requires lots of work (and food, and muscle, and bone tissue mechanical strength) when one is a quadruped on land. But it takes no work if one is an aquatic quadruped, because the animal's body density is nearly equal to the body density. This discovery adds physics in support of the view [37] that along its body-after-body parade of drawings of species from quadruped to biped (on the horizontal time line of Fig. 8.4), the intelligent ape had an "aquatic" relative that made the discovery that lifting itself is much easier while submersed.

The support from physics for the aquatic phase is stronger, and it speaks of aquatic movement. Body hair disappeared almost entirely not only because it was useless as thermal insulation in water (hair insulates when its interstices are filled with air [38]), but also because it was shaved off by friction against water during movement. This is why hair persisted in body regions that were not exposed to fast water flow during forward movement (head, armpits, pubic area, and chest area between and under the pectoral muscles). Standing tall in water was necessary for breathing above water, which is also why the *homo* has a longer neck and nose in comparison with other apes. The swimming is why the aquatic homo developed limbs (hips, shoulders, collarbones) pointing to the sides, like in a crocodile, not ventral like in other land quadrupeds. It is also why the aquatic homo acquired feet that look like the flippers of seals, without opposing big toes, which are clearly not for arboreal life. An added advantage of being bipedal came when the aquatic ape ventured on land, where it looked taller than it really was among competitors, before returning to the water for even greater safety.

Evolution is a macroscopic phenomenon. It is the universal tendency of the whole to morph hand in glove with its environment to flow more easily, for greater access. The "whole" is your system (the object) and its surroundings, not the system alone. The macroscopic reality of the evolution phenomenon is why the words "finite size" appear in the 1996 statement of the constructal law (cf. Chap. 1, p. 5).

Object and finite size are old concepts in human thought, much older than the mirage of the infinitesimal size. I stressed this observation in the very last paragraph of my 2000 book [11]; this way: "Most of this work (i.e. constructal law) could have been done two centuries ago, before thermodynamics. It is a mystery that this was not done then. Instead, modern physics embarked on a course tailored to the principle of infinitesimal local effects. Constructal theory is a jolt the other way, a means to rationalize *macroscopic* features, objective, and behavior (i.e. evolutionary design)."

Evolution means a lot more than change over time. It is much more valuable to understand and to anticipate evolution than to discern the change after the change happens. It is valuable because most of us think in terms of what is important to us.

In evolution, there is no reset button, no return to past success. The design is constantly evolving. The future design must be different than the present design. The environment is dynamic, always changing and morphing. Accept it the way it is, and design with it. Do not force the environment to be steady (static) and peaceful, or you will be defeated.

References

1. A. Bejan, S. Lorente, The constructal law and the evolution of design in nature. Phys. Life Rev. **8**(3), 209–240 (2011)
2. T. Basak, The law of life: the bridge between physics and biology. Phys. Life Rev. **8**, 249–252 (2011)
3. A.F. Miguel, The physics principle of the generation of flow configuration. Phys. Life Rev. **8**(3), 243–244 (2011)
4. A.H. Reis, Design in nature, and the laws of physics. Phys. Life Rev. **8**(3), 255–256 (2011)
5. L. Wang, Universality of design and its evolution. Phys. Life Rev. **8**(3), 257–258 (2011)
6. Y. Ventikos, The importance of the constructal framework in understanding and eventually replicating structure in tissue. Phys. Life Rev. **8**(3), 241–242 (2011)
7. G. Lorenzini, C. Biserni, The Constructal law: from design in nature to social dynamics and wealth as physics. Phys. Life Rev. **8**(3), 259–260 (2011)
8. L.A.O. Rocha, Constructal law: from law of physics to applications and conferences. Phys. Life Rev. **8**(3), 245–246 (2011)
9. J.P. Meyer, Constructal law in technology, thermofluid and energy systems, and in design education. Phys. Life Rev. **8**(3), 247–248 (2011)
10. J.A. Tuhtan, Go with the flow: connecting energy demand, hydropower, and fish using constructal theory. Phys. Life Rev. **8**(3), 253–254 (2011)
11. A. Bejan, *Shape and Structure, from Engineering to Nature* (Cambridge University Press, Cambridge, UK, 2000)
12. A. Bejan, J.H. Marden, Unifying constructal theory for scale effects in running, swimming and flying. J. Exp. Biol. **209**, 238–248 (2006)
13. J.D. Charles, A. Bejan, The evolution of speed, size and shape in modern athletics. J. Exp. Biol. **212**, 2419–2425 (2009)
14. A. Bejan, E.C. Jones, J.D. Charles, The evolution of speed in athletics: why the fastest runners are black and swimmers white. Int. J. Des. Nat. Ecodynamics **5**(3), 199–211 (2010)
15. J.D. Charles, A. Bejan, The evolution of long distance running and swimming. Int. J. Des. Nat. Ecodynamics **8**, 17–28 (2013)
16. R.E. Horton, Drainage basin characteristics. EOS Trans. AGU **13**, 350–361 (1932)
17. J.T. Hack, Studies of longitudinal profiles in Virginia and Maryland, USGS Professional Papers 294-B, Washington DC (1957), pp. 46–97
18. M.A. Melton, Correlation structure of morphometric properties of drainage systems and their controlling agents. J. Geol. **66**, 35–56 (1958)
19. A. Bejan, S. Lorente, A.F. Miguel, A.H. Reis, Constructal theory of distribution of river sizes, Section 13.5 in *Advanced Engineering Thermodynamics*, ed. by A. Bejan, 3rd edn. (Wiley, Hoboken, 2006)
20. A. Bejan, Street network theory of organization in nature. J. Adv. Transp. **30**(2), 85–107 (1996)
21. A. Bejan, Constructal-theory network of conducting paths for cooling a heat generating volume. Int. J. Heat Mass Transf. **40**, 799–816 (1997). (Published on 1 November 1996)
22. A. Bejan, Constructal tree network for fluid flow between a finite-size volume and one source or sink. Revue Générale de Thermique **36**, 592–604 (1997)
23. S. Kim, S. Lorente, A. Bejan, W. Miller, J. Morse, The emergence of vascular design in three dimensions. J. Appl. Phys. **103**, 123511 (2008)
24. C.R. Carrigan, J.C. Eichelberger, Zoning of magmas by viscosity in volcanic conduits. Nature **343**, 248–251 (1990)
25. A. Bejan, Rolling stones and turbulent eddies: why the bigger live longer and travel farther. Sci. Rep. **6**, 21445 (2016)
26. A. Bejan, J.D. Charles, S. Lorente, The evolution of airplanes. J. Appl. Phys. **116**, 044901 (2014)
27. R. Chen, C.Y. Wen, S. Lorente, A. Bejan, The evolution of helicopters. J. Appl. Phys. **120**, 014901 (2016)

28. A. Bejan, J.D. Charles, S. Lorente, E.H. Dowell, Evolution of airplanes, and What price speed? AIAA J. **54**(3), 1116–1119 (2016)
29. A. Bejan, R.W. Wagstaff, The physics origin of the hierarchy of bodies in space. J. Appl. Phys. **119**, 094901 (2016)
30. G. Ellis, J. Silk, Defend the integrity of physics. *Nature* **516**, 321–332 (2014)
31. A. Bejan, S. Ziaei, S. Lorente, Evolution: why all plumes and jets evolve to round cross sections. Sci. Rep. **4**, 4730 (2014)
32. A. Bejan, *Advanced Engineering Thermodynamics*, 2nd, 3rd, 4th edns. (Wiley, New York, 1997)
33. A.H. Reis, A.F. Miguel, A. Bejan, Constructal theory of particle agglomeration and design of air-cleaning devices. J. Phys. D Appl. Phys. **39**, 2311–2318 (2006)
34. A. Bejan, Why the bigger live longer and travel farther: animals, vehicles, rivers and the winds. Sci. Rep. **2**, 594 (2012)
35. A.H. Reis, A.F. Miguel, M. Aydin, Constructal theory of flow architecture of the lungs. Med. Phys. **31**, 1135–1140 (2004)
36. A. Bejan, *The Physics of Life: The Evolution of Everything* (St. Martin's Press, New York, 2016)
37. Elaine Morgan says we evolved from aquatic apes, TED 31 July 2009. https://www.youtube.com/watch?v=gwPoM7lGYHw
38. A. Bejan, Theory of heat transfer from a surface covered with hair. J. Heat Transf. **112**, 662–667 (1990)

Chapter 10
Diminishing Returns

If evolution is so ongoing and everywhere, why do so many things look as if they are stuck in time? The cat and the dog, the milk and the four limbs, they are the same as they ever were. Or, so it seems. Even in technology evolution, which is happening on a much shorter time scale, we see the ossification of form. The pencil, the fork, and the four wheels of the ox cart and the auto have not changed. Why?

The reason is the phenomenon of diminishing returns, which is as much a part of physics as freedom, economies of scale, hierarchy, and evolution. Diminishing returns are observed in freely evolving flow architectures that have become "mature." At the mature stage in the evolutionary design, the new changes that continue to occur have marginal or imperceptible effect on the broad outlook and performance of the whole flow architecture.

I begin with two illustrations of the meaning and value of this subtle aspect of the physics of evolution and freedom. These examples are from engineering design, although as we reach in conclusion (p. 134) the phenomenon rules all evolution, including social organization.

First, imagine that you have the freedom to shape a tube through which water is pumped to flow. The tube is straight, its length is fixed, but the shape of the tube cross section is free to vary. The opportunity to shape the cross section is the freedom. The possible cross sections are infinite in number: slits, wedge shapes, trapezoidal shapes, curvilinear polygons, and many more.

As you morph the shape in your imagination, you make a first discovery: the sharp corners of the cross section are bad features. They pinch the flow and act as concentrators of fluid friction. Corners that are more open are beneficial features. You arrive face to face with the secret: the possible designs are infinite; however, in this mountain are hidden a smaller number of cross-sectional shapes that stand above the rest. These are the cross sections shaped as regular polygons.

The regular polygon has a measure, and it is the number of sides, n. This is also the measure of freedom in how the drawing morphs. While drawing a hexagon, the eye and the hand move more freely (and in more directions) on the page than while

A. Bejan, *Freedom and Evolution*,
https://doi.org/10.1007/978-3-030-34009-4_10

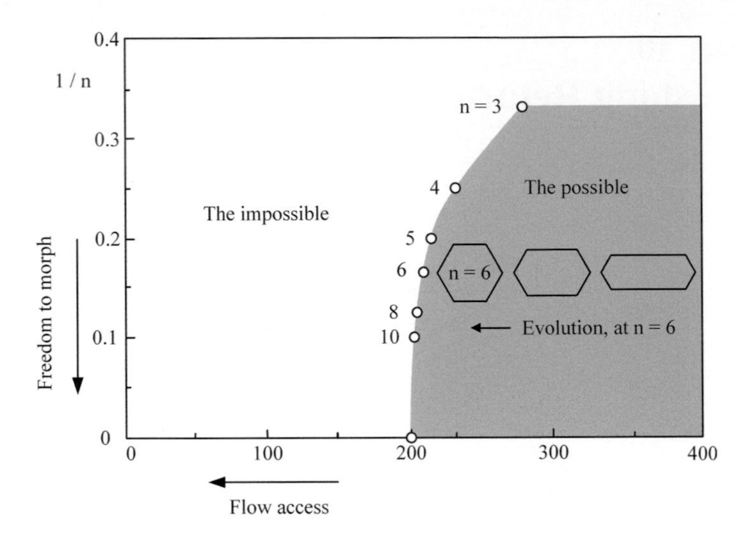

Fig. 10.1 The possible duct flow designs (right) and the impossible designs (left): how the flow through a duct finds easier access by means of greater freedom in morphing its design

drawing a triangle. Freedom increases as n increases, which is the arrow pointing downward on the ordinate of Fig. 10.1.

The numbers scribed on the abscissa are proportional to the pumping power required to push the water flow through the tube. The same numbers are also proportional to the pressure drop along the tube, the overall flow resistance posed by the tube, and the power destroyed while maintaining the flow. These details are available in Refs. [2, 5]. Here, important is the arrow that points to the left on the abscissa: easier flow access is available to the left, where the abscissa numbers are smaller.

Viewed from above, Fig. 10.1 presents a bird's-eye view of the movie of evolutionary design. The design must evolve toward the lower left corner. The rectangular frame represents the "design space," or the territory populated by diverse designs. As you search for easier access, your mind's images migrate toward the lower left corner. How, by injecting more freedom (n) in the image, unwittingly.

The secret is now in plain view. The few shapes that first came to mind are the regular polygons that do not have unusually tight corners: they form a string of points in Fig. 10.1. With more freedom, these pearls offer greater access. With infinite freedom ($n \rightarrow \infty$, or $1/n \rightarrow 0$), the flow cross section would offer maximum flow access, except that this shape is relevant as a theoretical limit, as a direction to the evolutionary changes that happen. The purely mathematical circle is not a physical flow cross section in nature or in ducts made by humans, because of imperfections and the unpredictability of the surroundings. In any case, the pearls are ordered on the string the way that makes sense to everybody. The round tube is better than other tube shapes.

Better does not mean much better. Diminishing returns is the phenomenon that rules. The flow access offered by a hexagonal or square cross section is not much

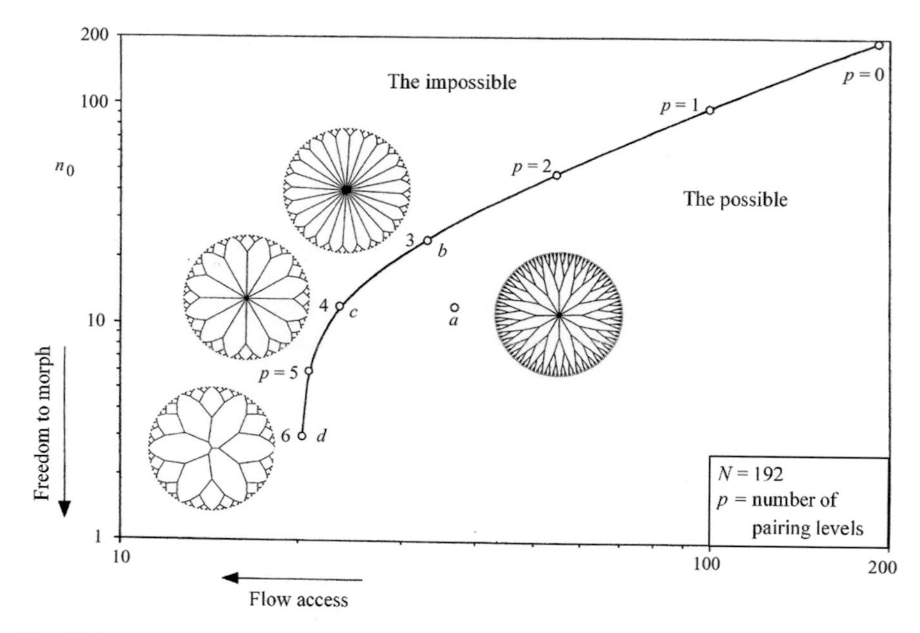

Fig. 10.2 The possible point-circle flow designs versus the impossible: how the vasculature finds easier flow access because of greater freedom in morphing its configuration

worse than the access through a circular cross section. The entire string is populated by winners. Yet, they all look different. They are diverse, but their performance is nearly the same. They are like the winners in the 100 m sprint, who every year climb on the podium with different faces, names, and flags. That's diversity. The faces are different, but the speed record is essentially the same. That's diminishing returns.

Diversity is natural, ubiquitous, and the reason why the trivial view of nature is that it is complicated and random. The subtle is the orientation of the string in Fig. 10.1, and the direction toward more freedom, which the string indicates.

The subtle becomes a loud a-ha! when we recognize that the string divides the design space into two worlds—to the right the possible, to the left the impossible. The possible is easy to imagine if we read the figure from right to left at constant n, for example, at $n = 6$. Constant n means cross-sectional shapes that are polygons with six sides. The point on the string represents the regular hexagon. To the right, on the same $n = 6$ line, are all the imaginable cross-sectional shapes with six sides that are unequal. None of those shapes perform as well as the regular hexagon. From the $n = 6$ population, the regular hexagon is the winner.

The impossible is unmasked by continuing to read Fig. 10.1 to the left, at constant n. No drawing can exist to the left of the regular hexagon. Looking for design to the left of the string is like looking for a flowing river basin inside a brick. You are powerful because you know where not to waste your time digging. The string of pearls that marks the limit of the possible is the secret of the impossible.

The second example is in Fig. 10.2. Imagine a vascular flow architecture that connects the center of a disc with N points spread equidistantly on its perimeter. The vasculature is composed of many branching tubes, each with laminar flow through it, like the flow of blood through the capillaries. The tubes are thinner as we move from the center toward the perimeter. At every branching point, the diameter of the branch is smaller (by a factor of $2^{-1/3}$) than the diameter of the mother tube. All these details are available in the literature [2, 5].

Important is that every possible design is represented by one point in the rectangular frame of freedom versus access. Note that Fig. 10.2 is the same frame as in Fig. 10.1, which is why this second example reinforces the conclusions drawn from the first. On the abscissa, the flow access between the center point and the perimeter increases toward the left. The numbers scribed on the abscissa are proportional to the pumping power and the resistance to flow between center and perimeter.

The freedom to construct the vasculature is measured in terms of the complexity of the drawing, which in Fig. 10.2 increases downward on the ordinate. Again, freedom is measurable. You can count the number of degrees of freedom that one has in making the drawing. The numbers indicate the value of n_0, which is the number of tubes that issue from the center. This number decreases when the number of branching levels (or pairing levels, p) increases at constant N. For example, the drawing in the lower left corner has six branching levels, starting with the three tubes coming out of the center, such that the number of points of the circle is $N = n_0 2^p$, namely, $192 = 3 \cdot 2^6$.

The possible designs are infinite in number. A child can draw something to connect the center of a disc with N points on the perimeter. More clever but equally arbitrary is the so-called "fractal" design labeled "a" and pictured at the same level with "c", on the horizontal line $p = 4$. Fractal design is a misnomer. It should be called "postulated (assumed) tree-like." This design is easy to draw by assuming that the length of each branch is a certain (fixed) fraction of the length of its mother tube. The constancy of the assumed branch/mother ratio of lengths is evident in the design displayed to the right of point "a".

Most difficult is to let all the tube lengths vary freely, and to conduct the minutious search for an architecture that offers progressively greatest flow access when N and p are specified. Three such drawings are presented in Fig. 10.2. In total, seven of these designs are represented, from $p = 6$ in the lower left corner, to $p = 0$ in the upper right corner. The $p = 0$ design has no branchings, just 192 tubes placed radially from the center to the perimeter. The $p = 6$ design has the most freedom to morph (the most numerous features to draw) and, as a consequence, it offers the easiest point-circle flow access.

All the possible designs fall to the right of the maximum freedom designs such as points "d", "c", and "b". The hand-picked design "a" falls well to the right of "c". We conclude that the exhaustive search for morphing the vasculature for progressively greater access ends up dividing the freedom-access space into two design domains. To the right of the winners indicated as points $p = 6, 5, 4\ldots$ is the domain of the possible. To the left is the domain of the impossible. The frontier between the two is the narrow group of designs that are most valuable.

The similarities between Figs. 10.1 and 10.2 are striking. In both figures, the frontier between the possible and the impossible has the same shape. The frontier

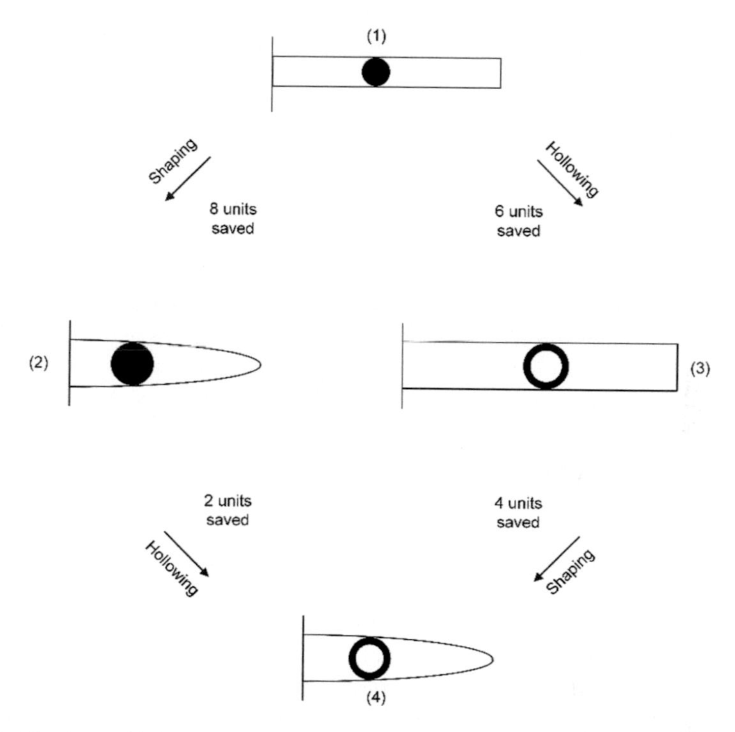

Fig. 10.3 Shaping and hollowing: two directions in the evolution of an elastic cantilever beam with freedom to morph. The load P is attached at the free end, and it points downward. The downward deflection of the free end (δ) is proportional to P

is not a continuous curve: it is an imagined curve drawn through the few points that represent the winners. The curvature of the curve indicates that "diminishing returns" become reality as the freedom and the complexity of the design increase. The performance of design $p = 6$ (point "d") is not much better than the performance of design $p = 5$. We reached a similar conclusion in Fig. 10.1, after comparing the hexagonal tube ($n = 6$) with the circular tube ($n = \infty$).

Diminishing returns is a universal phenomenon that accompanies evolution. The more mature the evolving flow configuration, the smaller the improvements in its overall flow-access performance. When the evolving "animal" is old enough the improvements are imperceptible, to the point that the observer believes that evolution has ended. The observer is mistaken. Evolution does not end, it just waits to be kick-started by its unruly environment into a new direction.

Figure 10.3 illustrates the diminishing returns phenomenon with an evolutionary design that facilitates "the flow of stresses" [5, 1] through a solid body that transmits a load. The cantilever beam is not usually thought of as a flow system, because it is a piece of solid material, yet, a flow system it is. The load P placed at the free end is transmitted along the beam to be felt in the armpit, where the beam is implanted in the vertical wall. The load is transmitted by means of stresses, which fill the body of the beam.

The cantilever beam is an artifact, one of the earliest contrivances inherited from before antiquity. The following description uses the language of mechanics, although the subject applies equally to the evolutionary design of the limbs of trees and the bones of animals [1, 3].

First, the beam happens because it has purpose. The beam improves the life, movement, and survivability of the greater system that adopts it. The greater system is the living human and the life of the whole society, in motion. In Fig. 10.3, the purpose of the beam is to support the end load without breaking and without bending too much. Resistance to breaking means that the internal stresses must not exceed a maximum allowable stress level (s_{ma}), which is a property of the beam material. Next, not bending too much means that the beam must have a certain stiffness, which is accounted for by a specified downward deflection of its tip, δ. The cantilever beam is essentially an elastic spring, a blade with a specified spring constant equal to P/δ. The greater the ratio P/δ, the stiffer the beam.

Imagine one of the simplest beam designs, the solid rod with round cross section shown as design (1) in Fig. 10.3. The analysis of a beam with specified P, δ, s_{ma}, and E (Young's modulus of elasticity) is relatively simple and known as slender-beam pure bending theory [2]. We can skip these details and retain the measure of the overall performance of design (1), which is the size of the beam, i.e., the amount of material that one must purchase and use to construct the beam. It turns out that the required volume of material is $V_1 = 12$ units, where the volume unit is the value of the group $EP\delta/s_{ma}^2$, which is a specified constant because its four factors are specified.

Evolution begins as we search for designs that serve the same need (P, δ) while requiring less material. Such designs are accessible through a series of changes in beam morphology, each change having the effect of removing from the beam some material that is not stressed as highly as the highest stressed regions inside the beam. In design (1), the highest stresses occur in the two armpits at the wall, the dorsal (the back, in tension) and the ventral (the belly, in compression). The lowest stresses (zero, in fact) occur at the free end and along the centerline of the solid rod.

These observations reveal the two directions to freedom so that the beam design can evolve in order to perform its function with less material:

1. *Shaping* the beam by removing material from near the tip. This way the beam becomes tapered, thick at the base, and thin at the free end, like all tree branches.
2. *Hollowing* the beam, so that the solid rod is replaced by a tube, like the bone of a bird.

Next come the interesting results, which reveal quantitatively the diminishing returns phenomenon. We imagine three future designs, in this sequence:

If "shaping" is the only direction to morph the design then the resulting design is the tapered solid rod (2), for which the required volume is $V_2 = 4$ units. This represents dramatic savings of material, measured as 8 units of material saved relative to design (1).

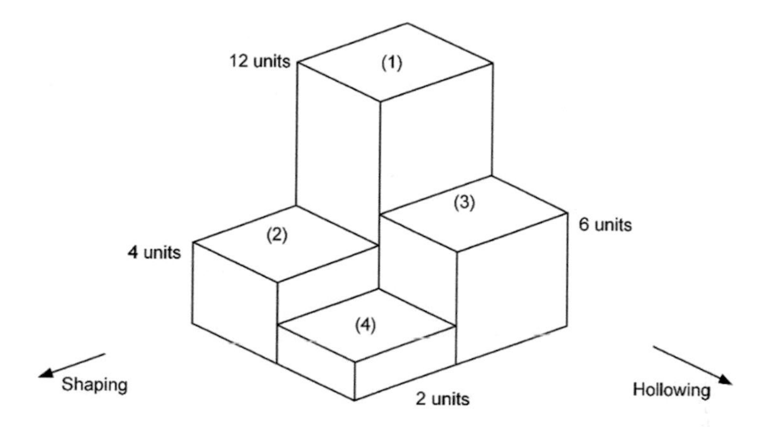

Fig. 10.4 Diminishing returns: the biggest savings in beam material occur early, when each design change is implemented for the first time, alone

If "hollowing" is the only way to change the design then the beam is a hollow tube with the volume $V_3 = 6$ units. The savings of material amount to 6 units, which are also dramatic.

Designs (2) and (3) are the "pioneering" designs, the first announcements of the invention that material can be saved by morphing the beam. Later, as the design matures the design evolution starts from either (2) or (3):

If the shaped design (2) is already available, then it is ripe to be subjected to hollowing. The next result then is a tapered beam that is also hollow, like the top of the original bic pen, or like the root of the goose feather. This is design (4), and its volume requirement is $V_4 = 2$ units. The change from V_2 to V_4 represents material savings amounting to only 2 units. Notice the diminishing returns that come when the design tends to maturity.

If, on the other hand, the hollowed design (3) is available for the restart of evolution, then it can be subjected to shaping. The resulting design is once again design (4), the hollow tapered beam, which represents the application of "shaping" and "hollowing" together. The change from V_3 to V_4 brings about material savings amounting to only 4 units.

Figure 10.4. summarizes the material savings identified so far. Note the downward steps in the volume of the beam. There are two directions for freedom in morphing the design, to the left is by shaping, and to the right is by hollowing. Time passes in the direction of the arrows shown in the figure. The evolving design matures in the direction of these arrows.

The biggest returns are the earliest, 8 units to the left and 6 units to the right, when the object (the beam) was untouched, and each of the design-change ideas was implemented alone. When the evolving design has become more mature, the returns are significantly smaller, 4 units and 2 units, respectively. Smaller steps happen when the two ideas are implemented together.

Diminishing returns become the norm as the design evolves by acquiring more and more design changes that proved to be beneficial in the past, when they were invented and implemented alone. They brought the greatest returns when they were new and not contaminated with similar ideas.

Diminishing returns are everywhere, most visibly in the human sphere, in the evolution of performance, in the shapes of boats, cars and airplanes, and the evolution of records in sports.

For example, ask why speed records are broken more frequently in swimming than in running [6]. The reason is that running is the "mature" design of human locomotion. We are born to walk and run—we are terrestrials. For modern humans, swimming is a young form of locomotion, something to be relearned after the prehistoric aquatic phase of human evolution (p. 119). Each swimmer must learn how to swim, to increase his or her access to movement on the globe, to avoid being stuck on one side of the river, and to gain access to the other side, which is greener. Running on soft sand and snow also requires learning.

Experts on one thing or another will surely jump in with other explanations for the difference between the frequency of records in swimming and the frequency of records being broken in running. They might mention the changes in equipment (suits, shaved body, pool depth, water quality). This argument is correct, and it reinforces the explanation given in the preceding paragraph. The equipment technology for swimming is young and the equipment technology for running is mature. The equipment and the rules of the sport are more likely to change in swimming than in running. The swimming pool is more likely to be improved than the track.

The reality of diminishing returns is the compounding of innovations, and it is rooted in physics. Old inventions are mature, full of sequential improvements that have become superimposed. These have very small returns from new design changes that are added on. Nowhere is this phenomenon more evident than in evolution of steam turbine power plants, which is a flow architecture dating from the late 1800 s. Although this flow architecture has become very complicated, in Fig. 10.5, we see two of the most essential design changes, reheating the steam and heating the water (the feed) before entering the boiler.

The secret to inventing a more efficient power plant is to morph the circuit executed by its working fluid (e.g., steam) such that it is at a higher temperature when being heated and at a lower temperature when being cooled. The more efficient design is the one that occupies a wider temperature gap between the heat source and the heat sink.

Two methods of widening this temperature gap are shown in Fig. 10.5. The stream of high temperature and high-pressure steam arrives from the left (from the boiler and a subsequent heat exchanger in the same fire house, called superheater) and flows through turbines that generate power. The steam expands, its pressure decreases, and so does its temperature. The design change consists of intercepting the stream halfway through the turbine, and heating it in a special heat exchanger called reheater, which is also exposed to the fire. This way the turbine becomes segmented into two turbines, one for high-pressure steam and the other for low-pressure steam, and the

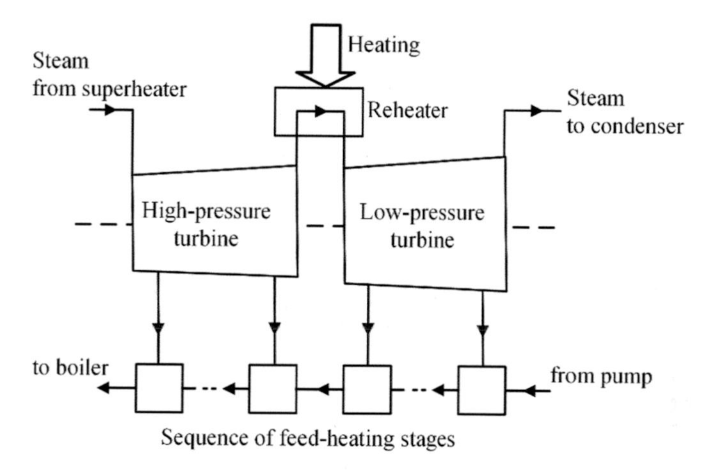

Fig. 10.5 Reheating the high-pressure steam flowing through the turbines of a power plant, and preheating with bled steam the liquid water before it is fed into the boiler

temperature of the steam (averaged over the two turbines) is higher than what it was before the reheater was invented.

Downstream of the low-pressure turbine the steam is condensed into liquid water in a heat exchanger (called condenser) exposed to the cold ambient. Next, the liquid flows through a pump, which increases the water pressure to the high level required for boiling water at a high temperature when exposed to the fire.

The lower part of the drawing in Fig. 10.5 shows another design change, which is for preheating the pressurized water stream arriving from the pump and flowing into the boiler. The smart way to heat the water fed to the boiler (called feedwater) is by placing the water in contact with steam bled from the turbines. This heating invention is valuable because it avoids the mistake of placing the cold water from the pump in direct contact with the fire. Heat transfer across huge temperature differences is a killer of efficiency. The thermodynamics term for this mistake is irreversibility or entropy generation. Again, technical terms and details can be skipped, and can be found in the literature [4].

Important is that in Fig. 10.5 there are two very good inventions, one is reheating and the other is feedwater heating. If implemented, each invention causes an increase in the efficiency of the whole power plant. The efficiency η is the ratio of the shaft power delivered by the turbines divided by the rate of heating (or the rate of fuel consumption) administered to the steam before expansion through the turbines.

Although both reheating and feed heating lead to improvements in the efficiency of the steam turbine cycle, the efficiency increase caused by one method is greater when the method is implemented for the first time, alone. Note the distinction between efficiency (η) and the efficiency increment ($\Delta\eta$) resulting from one or more design changes. The efficiency is the highest when reheating and feed heating are used at the same time.

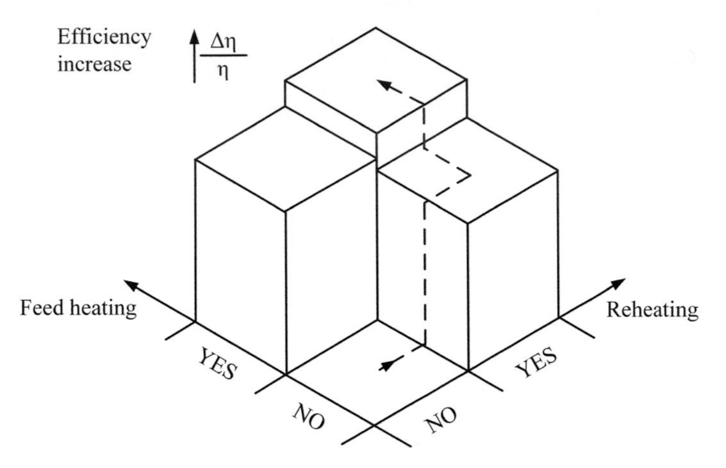

Fig. 10.6 Two design changes that lead to higher global efficiency: reheating and feed heating. The efficiency increase is most pronounced when each change is implemented alone. The increase is smaller when the two inventions are implemented together. Review the same message in Fig. 10.4

The impact of a single method depends on whether the other method was implemented already. This point is made graphically in Fig. 10.6, in which the relative efficiency increase $\Delta\eta/\eta$ is plotted in the vertical direction. Big returns are registered when the invention is applied first, by itself. Later, as the flow architecture is mature, decorated with beautiful features like a Christmas tree, diminishing returns come from subsequent increase in global performance.

There is a lot more to what I selected here for illustration. For example, the number of feed heaters (n, Fig. 10.5) is free to vary. Each feed heater is a flow system with its own flow architecture that can be morphed freely toward better performance, i.e., greater efficiency at the whole system level [4]. It turns out that the design with only one feed heater causes an efficiency increase of 9%, whereas the design with an infinite number of feed-heating stages causes an increase of 20.7%. In other words, the fresh invention (a single feed heater) delivers half of the benefits offered by the most mature and perfected version of the invention (continuous feed heating, or $n = \infty$).

It's never like the first time. Returns are like squeezing a lemon. The biggest squirt comes when you squeeze it the first time.

"Gilding the lily" is an apt metaphor for the evolutionary phenomenon that has become mature. Where there is freedom, evolution happens and along with diminishing returns in global performance it gives birth to diversity and complexity. This is, in a nutshell, the visual message of Figs. 10.1, 10.2, 10.3, 10.4, 10.5 and 10.6 and the physics that underpins them. Gilding the lily is most of what goes on in a mature science. It is a good thing, but only up to a point that announces its presence naturally, and when it does it changes the movie plot.

A bird's-eye view of the diversity and complexity that ensues is provided by Fig. 10.7, which shows how the body of designs for steam power plants has been

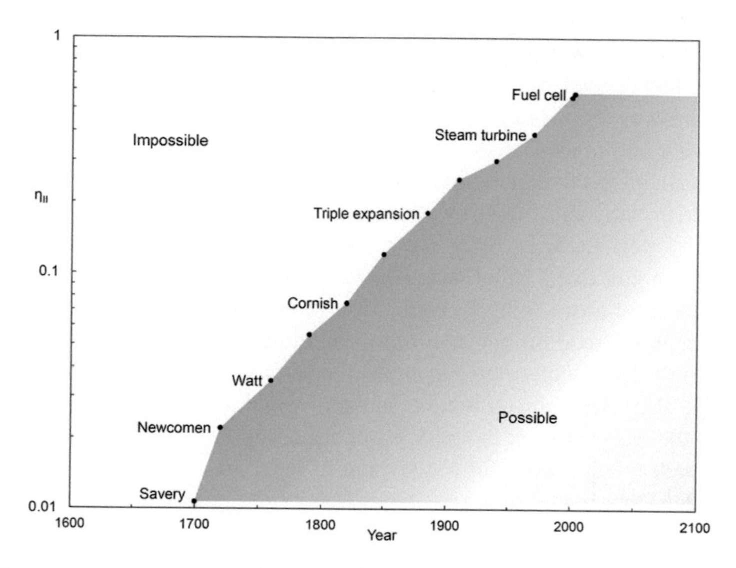

Fig. 10.7 Highlights in the evolution of the second law efficiencies of steam power plants during their history. The data that mark the crest of the domain of possible designs are from Ref. [4]. Note the logarithmic scale on the ordinate, and the $\eta_{II} = 1$ ceiling that no design can surpass. At every point in time, the crest divides the designs into the possible and the impossible, cf., Figs. 10.1 and 10.2

growing since the harnessing of power from fire 300 years ago. This volume is like a mountain, the crest of which is rising over time.

The crest is marked by pioneering designs that at the time were the most efficient. In other words, the shaded body of the mountain accounts for the multitude of designs that are less efficient than the designs that define the crest. On the ordinate, η_{II} is the second law efficiency of each design, which is a number that cannot be greater than 1. The second law efficiency is the ratio of the power output of the design divided by the corresponding power output of an ideal design (known as reversible engine, or Carnot engine). The crest of the mountain cannot push through the ceiling represented by $\eta_{II} = 1$. Consequently, the crest must be concave, and the concavity in this figure is an illustration of the phenomenon of diminishing returns in the evolution of mature flow architectures.

The subtle aspect that this science illuminates is what is *impossible*. This is what unites the examples shown here in Figs. 10.1, 10.2, 10.3, 10.4, 10.5, 10.6 and 10.7. This knowledge is very valuable. If we know it we do not waste our time touching and feeling our way over the cliff. This way we avoid the big mistakes. All science is about finding the limit between the possible and the impossible, and how to push the limit, if possible.

This conclusion applies broadly and reaches back to the physics of social organization, with which this book began. The unending evolutionary design of society is better known as politics—the proposals, commands, and implementations of changes in the regulations to human flow on the earth's surface such as the *polis* (the city, in

Greek). Changes happen, all the time. Some changes are big, while many are small. Some are sudden, cataclysmic, while most are tiny, slow, and imperceptible. Changes happen because every one of us has the urge to change something. Humans cover the range from the complacent to the revolted, with the many dissatisfied in between.

Otto von Bismarck famously said "Politics is the art of the possible." This wisdom is old, shared broadly and known by other names, such as realism, compromise, trade-off, flexibility, perfection is the enemy of performance, can't win them all, and the benefit of changing one's mind. More powerful and useful is to have this wisdom put on a scientific basis. Why, because "the possible" is essentially an infinity of possibilities. With freedom, the social flow architecture can be changed in all sorts of ways, minor and major, related to each other or not related. When it comes to the possible, the sky is the limit. If there is freedom, anything goes…until the morphing design hits the wall between the possible and the impossible. Without freedom, the design is stuck much sooner far to the right in Figs. 10.1, 10.2, and 10.7, far from the better ideas that would be accessible if freedom to question were encouraged.

The way forward is to become better educated about the impossible and to implement what is possible and free, economical, safe, robust, resilient, and long lasting. With science, every new generation is brought up with a greater ability to construct and predict its own future.

True is also that every new generation arrives soaked in unpredictability. Individuals and ideas cannot be predicted. They are like the eddies that rotate "the wrong way" on the surface of the flowing Danube. To focus on the individual is to be blind to the whole. The evolving flow design that serves all the eddies is the river basin, the whole. This is why the future of the whole is the important picture to paint. The better the science, the more clairvoyant and far seeing the painter.

References

1. A. Bejan, S. Lorente, J. Lee, Unifying constructal theory of tree roots, canopies and forests. J. Theor. Biol. **254**(3), 7 October 2008, pp. 529–540
2. A. Bejan, *Shape and Structure, from Engineering to Nature* (Cambridge University Press, New York, 2000)
3. A. Bejan, The constructal-law origin of the wheel, size, and skeleton in animal design. Am. J. Phys. **78**(7), 692–699 (2010)
4. A. Bejan, *Adv. Eng. Thermodyn.*, 4th edn. (Wiley, Hoboken, 2016)
5. A. Bejan, S. Lorente, *Design with Constructal Theory* (Wiley, Hoboken, 2008)
6. M. Futterman, Personal communication, July 20, 2017

Chapter 11
Science and Freedom

Freedom and organization are thoughts and happenings that unite us as members of civilized society. The physics is the universal phenomenon of evolution with its many familiar manifestations such as economies of scale, social organization, diminishing returns, the spreading of ideas, and the relentless generation and adoption of better flow architectures over time. This program arches back to science itself—science as an evolutionary and self-correcting add-on that empowers humans. Why is this important? For at least four reasons:

First, if you look at the world map today you see a physical flow that carries all the human features and concerns that we have discussed. The flow is organized nonuniformly, hierarchically on the earth's surface. You see the same hierarchical movement taking shape in your mind as you review the history lessons that you were given: books, movies, names, places, and human events covered in the press.

Second, it is always the case that the society moves more, produces more, and is longer lasting when it is endowed with freedom, hierarchy, free questioning, and self-correcting. These physical features are also distributed nonuniformly, in ways that are obvious. This second flow architecture coincides with the hierarchical distribution of creators and transmitters of science, on the same world map. This is no coincidence.

Third, freedom is a physical feature, such as shape, dimension, weight, change (process), and power. Freedom is measurable. Nothing changes, nothing moves, and nothing evolves unless it has freedom. The freedom property is a measurement of how many features are free to be changed in the flow system configuration. Measurable is also the physical effect of freedom on efficiency and performance. Freedom is also measured as the number of "degrees of freedom" that are present in the constitution of the model of the observed (physical, natural) configuration of the flow system. Degrees of freedom are those features that can be changed freely, independently of other features.

Evolution without freedom is nonsense, because one cannot have design in nature (live, morphing to flow more easily over time) without freedom to change. The straight steel pipe is not a live system because it does not have the freedom to morph,

A. Bejan, *Freedom and Evolution*,
https://doi.org/10.1007/978-3-030-34009-4_11

to improve its flowing in an evolutionary manner. The drawing of the steel pipe is dead. The turbulent water stream in the river channel, through the marsh, and even through the steel pipe, is a live flow system. It has configuration, freedom, evolution, and it persists in the future. In one word, it has "life," just like all the other flow configurations with freedom to change, from animal evolution to technology evolution and society evolution.

Nature behaves in the same way, imperceptibly, all the time, and on a much broader range of degrees of freedom. This is why with the constructal law we have been able to predict the designs of inanimate flow systems (river basins, turbulence, snowflakes) and animate flow systems (lungs, vegetation, animal locomotion, human, and machine evolution, e.g., aviation). We can use this method to investigate and innovate social, political, and technological systems as well.

Every tree canopy, branch, and leaf avoids the space of the neighbor because it must have free access to fresh air. The spacings between canopies appear to be carefully designed and sized. There are also carefully sized spacings between leaves. From this physics comes the arrangement of branches and leaves, and the few large and many small mosaic of tree canopies on the forest map (cf. Chap. 3).

The fourth reason why the physics of freedom is important is science itself. After all, what is science? That's easy, science is a kick. Why is there such a pleasure, this feeling of having a hunch, figuring it out, discovering that you are right and then telling everybody? The science question is about all of us. Why it is human to have hunches, to want to know and, if possible, to know in advance, to predict?

The answer is that all these urges—to have food, shelter, knowledge, and longevity—are design features that facilitate life, which is the movement of all animal mass over the earth [1–3]. Without such design, our mass would not be moving as easily and as far. Without other flow designs (physical streams, inputs called food, fuel, useful energy), our mass would not be moving at all.

Science lives up to its literal meaning, which is knowledge (*scientia* in Latin). As we come to know more, we become more reliant on our knowledge to do something with it, for us. We predict better what will happen if we make decision A, and what will happen if we make decision B. We compare the anticipated effects of A and B, and we choose. Knowledge is the ability to implement design changes that are useful to the individual and the group. With physics, we design the future, we predict it, we build it, and we walk into it. Intelligence is to "see" a better design before the better design is spoken, tested, and built. Intelligence is the fast-forwarding of design evolution.

Knowledge is science, and it evolves by observing, condensing, and streamlining mentally the flow of observations. The condensed are the principles, and, among them, the most unifying are the *first principles*, the laws. Few large and many small is the hierarchical flow design that empowers everything about us as we move on earth and in life. In the flow and evolution of knowledge, the few large are the laws, and the many small are the observations and the data.

Knowledge (science, information, news) flows on the globe because it is moved by moving individuals. It flows from individuals who move more (because they possess knowledge) to individuals who move less and have the need to possess knowledge. When both ends of this flow possess the knowledge, the flow of knowledge stops.

The spreading of knowledge is often described as the "diffusion" of information. Seen from physics, the term "diffusion" is not correct. The spreading of anything is a combination of two ways to flow. First, the spreading is by "invasion" by carriers along fast and long channels that stretch across the available territory or population. Second, the interstices between the channels are "consolidated" by diffusion perpendicular to channels, which is slow and covers short distances. The coexistence of the fast and the slow, or the long and the short, is why any spreading or collecting flow has an S-shaped history of how the covered territory increases in time.

Physics is not a rigid text in today's physics textbook. Science evolves, because we all want to predict the future. We are being selfish. We design this future to be good for us, with ourselves inserted in it. This virtual future walks and drives with us, in front of us, like the carrot in front of the horse, except that we have it better than the horse: we create food and fuel, more and more, and we keep going. In this future, we make choices, all the time. We go with the flow, and the flow goes with us. Without knowledge we would be crawling back into the caves, fearful of everything that moves. Read the world news, this still happens today.

I remember one morning when I was a sophomore, walking to class. The subject that day was one of the pillars of mechanics, called Strength of Materials, which is a misnomer for the resistance (stiffness) of loaded structures. I was reviewing in my mind the previous lecture, which had been about how to select the thickness of a steel bar such that it will not bend much when a specified weight was attached to its end, as in Fig. 10.3. When I was crossing the street I was struck by a mental image so powerful that it felt dangerous. I saw that by knowing the principles I knew what *will happen* to that bar. I knew the future of the bar and of those sitting on it.

The monumental value of science is subtle. Crossing the street I saw that I was being given the power to predict the future. Not one future but several, one for each bar that I was contemplating. Even better, I was being given the ability to select the future that serves us best: the future with a bar that is strongest, lightest, and easiest to build. This meant that I was being empowered to design the future. I, a veritable nobody from nowhere behind the Iron Curtain, was acquiring powers that before science were attributed to the work of God.

Good ideas travel far and persist. Science is a story that flows freely from those who know to the many more who wish to know. This flow spreads on the globe and in time. It flows from generation to generation. It flows one way, from high to low, from a source to a population on an entire area. The sources have been many. Among them, the Golden Age of Greece was the biggest jolt forward in science. The heat engine, the industrial revolution, the telephone, and the Internet are mere puffs of smoke from the engine of geometry and mechanics.

The effect of science is measurable in watts. The allocation of more power to more individuals is knowledge, which represents the design changes that keep us alive and flowing easier, moving farther, and lasting longer in time. Collaboration means working together, organization, life. With freedom, the flowing entities are free to change. They move to the right, and then to the left, and find better ways of flowing.

Organization and design happen naturally. The collaborators do not know that they are collaborating, and know even less about with whom they are collaborating. They just find themselves interlinked on the globe in ways that serve them well individually. Hierarchy is in all these flow architectures.

The science that we learn and teach today is compartmentalized and told in several languages. There is one language for experts on animate things (animals, plants), another for experts on inanimate things (the rivers and the winds), yet another for experts on "natural" things (the animate and the inanimate combined), and finally the language for "artificial" things (society, economics, engineering). This is how we are raised, and as a consequence we think that humans are not natural. This makes as much sense as thinking that humans do not obey the physics laws of mechanics and thermodynamics.

Nothing cuts and divides more deeply than language. The first duty of the scientist is to learn the languages in which the original ideas were first published, and the history of the scientist's discipline. The second duty is to teach languages and history to disciples. Start with French, Latin, Italian, and German, and then learn English. It pays to publish in English.

The organization of science is maintained and reinforced by its establishment, which is the few embedded in the very many, the nobodies, the upstarts. This is a natural hierarchy (cf., Chap. 3). It is difficult to question the established view, particularly when the questioned phenomenon is everywhere. The more common the phenomenon, the less likely it is that it will catch our attention. For example, throughout human history everybody knew that air has no weight. What could be more obvious than the fact that the balance at the market does not tilt unless one puts something on the platter? Obvious was also that the sun rises, passes over our heads, and sets on the horizon, and therefore the sun circles around us. Admit it, you grew up thinking this way. In fact, most of the people on earth today still think this way. This is as obvious as thinking that running, flying, and swimming are three different movements, and that gravity does not matter to the fish.

Can a nobody, an amateur who accidentally and completely innocently wanders on the wrong street, convince the marching crowd that there is a better idea? Yes, the amateur can do it, with freedom and a respectful disregard for what others say, and only by persisting.

Creativity, risk, and punishment go together in scientific work. Yet, there will always be lone rangers, the few who climb on a limb, fall, and climb again (Fig. 11.1). They are the givers, the innovators, the true altruists, and the gift that keeps giving.

In science, a truly original idea rocks the boat. The initial reaction is silence. The second is quiet skepticism. When the idea begins to spread the skepticism turns into attacks, and later into their complete opposite: adoption, claims that it was not new, and plagiarism.

Many see the obvious, fewer see the subtle. Some are at home in both, the obvious and the subtle. The subtle is the bird's-eye view, which empowers us to predict evolution in nature and it reveals the key role of freedom in making evolution possible. Nature may look complicated, but it is in fact a tapestry woven in a very simple old loom. The designs consist of many flow types and sizes, all governed by a succinct

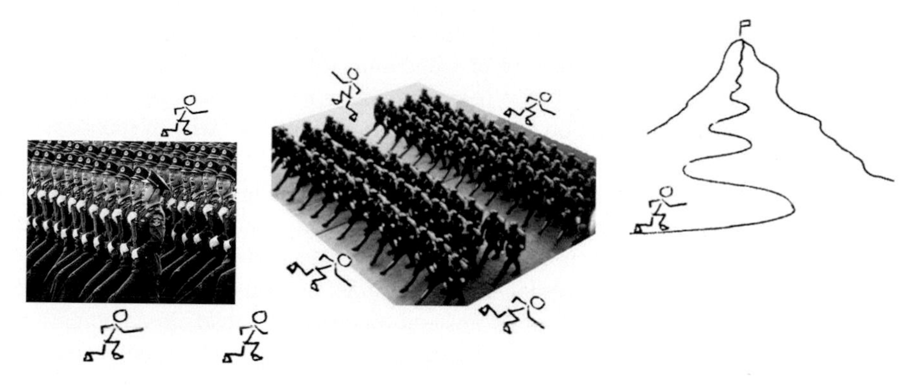

Fig. 11.1 Marching columns do not climb peaks

law of physics. All the designs fit because they flow together with their environment, the animate and the inanimate, the small and the large, and the human and the not human. They do not fit perfectly, and never will.

The more we rise above the details, the simpler the tapestry of nature looks. Taking a bird's-eye view is very good medicine for those dizzied by the smoke that nature is complicated, diverse, random, infinitesimal, nondeterministic, fractal, turbulent, nonlinear, and chaotic.

I learned the bird's-eye view method at MIT in the classrooms of Building 3, where I also learned my English. I remember the words, the times, the places, and the professors who taught me key words. I first heard the expression "the bird's-eye view" with a Dutch accent, from J. P. Den Hartog, my famous professor of dynamics. He was an artist of the simple, in a discipline that was already cluttered with many complex mechanisms. That was decades before the blur of computer-generated simulations of anything today. He taught us to step back, look at the whole, make it simple but and not "throw the baby away with the bathwater." Decades later, while lecturing in Amsterdam, I discovered that the baby thrown out with the bathwater is a Dutch saying. Professor Den Hartog was teaching the art of seeing the essential, and keeping it simple.

Luckily, at the same time, I was learning the bird's-eye view as a method in thermodynamics [1–4]. This is how thermodynamics was being taught at MIT in 1969, as a method to think, in addition to how to perform accurate calculations that keep the world warm, moving, efficient, wealthy, and safe.

The bird's-eye view method in thermodynamics is called "the control volume," which is an imaginary bag in which all the components and fine details fit. The control-volume method is a provocation to the student: if, as you claim, the components of your system obey the laws of thermodynamics, then show us that your whole ensemble of components obeys the same laws.

Very often the student fails this test. No, not because of lack of training. The student fails precisely because of the training, which is *reductionist* and pushes the mind in the direction of analyzing smaller and smaller parts. This training instills the belief that the truth lies hidden in the infinitesimal. Statistical mechanics grew along that path. So prevalent is the doctrine of reductionism that most of my peers think that something is "fundamental" if it is small enough to be present in everything. This thought is wrong, a violation of language, and certainly not why the concept of fundamental is useful.

Look at the edifices of ancient Rome, which were significantly taller and stronger than in earlier periods elsewhere. Sure, tiny packets of clay are found throughout the Roman edifices, in every corner of every brick. But, clay was present in everything else before Rome. The fundamental in the buildings of Rome was the use of the *fired* brick and cement, which could withstand much greater loads in compression and bending than earlier building blocks made of dried mud. The fundamental is the building block. The brick is macroscopic, not infinitesimal. The fired brick and science (architecture, mechanics) are why Roman buildings were a new "species" among buildings, taller and stronger, and with much bigger vaults and air spaces.

Fundamental means deep down, at bottom, not infinitesimal. The "fundamental" is the truth that lies as "foundation" (from the Latin noun *fundus*, which means bottom). The secret—the brick—is of finite size, not infinitesimal. The difference between finite size and infinitesimal is like the difference between black and white, night and day, dead and alive, and pregnant and not pregnant. The infinitesimal does not have freedom, flow, organization, and evolution. The finite size does.

There are fundamentals hidden in *applied* physics, and they deserve to be discovered. There are "idea people" who are not scientists, and they deserve to be heard. All science is useful. With science we know more, we remember less, we have less work to do, and consequently we have more time to live, learn, and create.

Freedom is the mother of all evolution and science. If you doubt that, think of it in the opposite direction, to the absurd: What kind of science would that be where the choices made long ago are already the best, rigid forever? It would be useless, with no purpose and no future. It would be the opposite of the science that attracts us, inspires us, and empowers us on earth.

Science is here to be questioned. When science becomes the authority—invoked and implemented by the state, as religion was many centuries ago, then science will give way to a new form of human pursuit of truth in nature, just as religion did. We see new forms already, all accentuated during the age of the Internet: independent scientists, self-publishing, predatory journals, false science, stealing credit, plagiarism, science journalism that overshadows and overpowers science publishing, and journalists who are far more powerful than the scientists whom they parrot.

> Journalism, you see, is the religion of modern society.
>
> (Honoré de Balzac, La peau du chagrin)

The wonderful thing about the human mind is that it has the natural urge to rationalize the observed, and vice versa, to observe in order to reason and put the observation of nature to good use, to empower man, family, and offspring. The mind springs into action. The mind of the listener does not check the title printed on the speaker's diploma. Anything goes.

Centuries ago scientists knew this because they themselves had begun as "amateurs," which means "lovers" of what they were doing. They discovered and invented because they were curious, not because they were digging for usefulness, applicability, practical importance, and pleasing the sponsor.

Today, in the knowledge industry, most scientists are unaware that the most important ideas that they cherish came originally from amateurs, from nobodies who were just curious. This is true across the board, not just in science. Opposite the amateur, every established scientist has an axe to grind. Any peer review written by an establishment scientist should be taken with a grain of salt.

Scientific revolutions do not happen through the accumulation of new data. Revolutions happen while looking at existing data and seeing their organization—their message—in a new way, suddenly, involuntarily, accidentally, unwittingly. Over time, the science improved by revolution tends to usurp the authority of academicians and politicians. Such changes are much slower under totalitarianism where false science and groupthink are raised to the level of truth. Examples of this kind are genetic theory in communist USSR [5] and the alleged analogy between heat transfer (heating a solid body) and work transfer (charging an electric capacitor) in communist P. R. China, which brazenly violates the laws of thermodynamics (reviewed in [1–4]) (cf., p. 143). Much of this is nationalism in the guise of scholarship [6, 7]. False science [6–12] joins the fake blue jeans, fake iPhones, and pirated books [13].

Many in academia make the mistake of equating the goodness of an idea with the number of authors who agree with that idea. The history of science proves this to be false. Science is not about counting people. All individuals are not equally imaginative. All opinions are not equally important. Science is not democracy.

Many make the mistake of equating the goodness of ideas with the volume of research funding, people hired, money spent, and buildings built. The history of science proves this to be false. Science is not the amount of money spent. Dollars spent are not equally important. When I read the scientific literature, I see names, dates, and ideas, not budgets. Science is not accounting.

When I read the news of a huge research grant for a big research center, I predict that it will change nothing. The truly new, the svelte, beautiful, and valuable comes from the least expected source, which tends to be the nobody with a zero budget. That's science again, and it is just like in competitive athletics: you have no idea what poor kid walks from the street onto the playing field. That is wonderful, it is the good news that sustains science and civilization.

Not everything about science today is rosy. An old habit threatens the disinterested truths that serve as support structure for science. Redoing a published idea, disguising it, and publishing it as "novel" is a dangerous trend, and it is rampant [14–20]. The National Science Foundation (U.S.) defined this kind of academic misconduct this way (NSF-CFR-689):

> Plagiarism means the appropriation of another person's ideas, processes, results or words without giving appropriate credit.

Because of electronic publishing and related enhancements in the speed and territorial reach of the flow of science, journals have ballooned in volume and number.

Compared with how science was created and transmitted when I was a student, today it seems everybody writes and nobody reads. It has become much easier to cheat, and much more difficult to police. The cheaters get away with it because administrators of our institutions (universities, journals) are not affected: plagiarists do not steal from individuals who published nothing worth stealing.

> Corruption is in force, talent is rare. Thus, corruption is the weapon of the mediocrity that abounds.
>
> (Honoré de Balzac, Le père Goriot)

The main purpose of science is to live a happy, creative, and long life, without the difficulties that create unhappiness, despair, giving up, and early death. In this big river, one benefits from swimming not alone (cf., Chap. 2). Yet, freedom is key: it pays to be independent, uncompromised, not a joiner.

All research is autobiographical. It is a human story about the researcher, the author, the people close to the author, the moment, the place, the language, and the excitement. I discovered this in my early teens when for Christmas my mother, the pharmacist, gave me Max von Laue's science story "The History of Physics". I still have this book. Likewise, the story of the present book was about ideas and human events in science. The story behind the story is how science "happens," how it evolves, and why science is good for all of us, so good that we keep telling its story. Think of science as a good joke, in fact, it is the best joke because it is repeated the most.

Here is a concrete example, which will probably make you laugh. Repeated misunderstandings that have long been corrected in science have a lot in common with repeated claims of inventions of *perpetual motion* machines [2]. The history of such inventions shows what to do about those who keep repeating claims that are known to be untrue. Two hundred years ago, the *Institut de France* adopted the policy of not accepting any more claims of perpetual motion inventions because (1) they did not work and (2) they had been proven invalid based on the old science of mechanics. A clock cannot turn forever, because of friction. To deny review and publication to such claims was not censorship then, and it is not today. On the contrary, it is liberating and encouraging the evolution of science.

What is to be done? The answer is obvious: Stop publishing and sponsoring falsehoods, teach the disciplines correctly, improve them along the way, make them even more general and powerful, and clean up the misconceptions that tend to arise during the evolutionary design of science.

Just look at the perpetual motion machine idea claimed in Fig. 11.2, which was sent to me by an observant colleague in the 1980 s. I found it so ingenious that I used it as a teaching opportunity ever since [2, 4]. What can be more obvious and appealing than this device for the production of mechanical power $\dot{W}$ $[watts]$ solely from the atmosphere of temperature T_0 [K]? Fuel is not needed. The air stream $\dot{m}$ $[kg/s]$ is used first as heat source in the heater (or boiler) of the power plant. The temperature of the air stream drops as it flows through the heater; therefore, the stream is used next as heat sink while flowing through the cooler (or condenser) of the power plant. The air stream is finally discharged into the atmosphere. What's wrong with that?

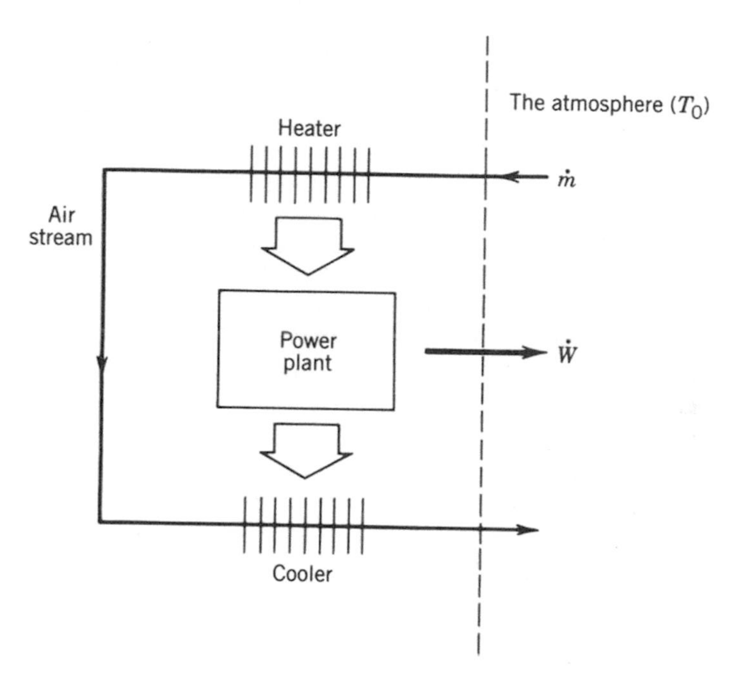

Fig. 11.2 A method for the production of mechanical power ($\dot{W}$) from the atmosphere, T_0. The air stream ($\dot{m}$) is used first as source in the heater of the actual power plant. The temperature of the airstream drops as it passes through the heater; therefore, the stream is used next as heat sink while flowing through the cooler at the power plant. The spent air is later discharged into the atmosphere. Is this design possible?

Wrong is the focus on the particular scheme, which is the inventor's claim. Wrong is to lose sight of the big picture, the holistic view. It's like watching the hands of the magician, not the whole stage. At fault is modern science education, which is *reductionist*, and this is particularly the case in physics. The cure is to step away from Fig. 11.2 and recognize that to the right of the dashed line the air stream loop must complete itself while in thermal contact with the ambient. The whole system that contains this complete drawing is a closed system in steady state, with an impermeable boundary of one temperature (T_0) and with the atmosphere (and the supposed recipient of $\dot{W}$) as its environment. Inside this closed system, the air stream completes a loop.

For the whole system, the first law of thermodynamics requires that $\dot{W} = \dot{Q}_0$, where $\dot{Q}_0$ is the heat transfer rate between system and environment. If the inventor is right, then power is produced ($\dot{W} > 0$, i.e., power leaves the closed system) and heat must be transferred from the atmosphere to the system ($\dot{Q} > 0$, which means that heat transfer enters the system). These flow directions (heat in, work out) violate the second law of thermodynamics for this class of systems: closed, thermal contact with only one temperature reservoir (T_0), and operating in steady state or an integral number of cycles.

We have rediscovered here that, at best, the "whole" can be a purely dissipative system that converts work into heat, one way, like any brake or clock mechanism. In other words, $\dot{W} < 0$ and $\dot{Q} < 0$, or work in and heat out. The inequality sign matters. This proves mercilessly that heat transfer is not analogous to work transfer, and that publications that continue to push false physics are worthless. The heat–work analogy is an error of the same rank as a perpetual motion invention.

The evolution of the physics of power (thermodynamics) shows that what works is kept as an add-on to the science that was [1–3]. What is false is swept aside, and forgotten. This is the evolutionary morphing design of science. This is also why every once in a while the scientific community is presented with a reality check, a new bird's-eye view that is suddenly useful as a guide to the new generations. Revisionism is checked, authority is questioned, fakes are exposed, mistakes are corrected, and this way a renewed appreciation of the discipline empowers the new generations. This happens sometimes in a spontaneous paper, a new perspective, a new review, and a new book. Researchers, authors, university administrators, national academies, publishers, and especially editors learn from this. The stream flows better after the rotting tree log is swept out of the way.

> Beware of false knowledge; it is more dangerous than ignorance.
>
> George Bernard Shaw

Science is like a civilized territory that improves, prospers, and expands because it makes life better for the people who belong to it. It expands as long as it keeps producing useful things, which attract people. The civilized welcome the newcomers—the nobodies—provided that they obey the laws, the disciplines. People join because their lives become better that way.

To fight the barbarians who pillage on the perimeter is a necessary and unpleasant effort, a nuisance, not the objective. As the defeated are assimilated and civilized, the civilized territory expands and, as a result, life, peace, movement, and freedom flourish.

The civilized territory that does not fight the barbarians is destined to disappear along with the good way of life that it was sustaining. The village without dogs falls prey to the wolves. Science, like all the useful artifacts produced by the civilized, is no different.

Mistakes will continue to happen. As the paleontologist Michael Taylor [21] noted, "Science is not always right—very far from it. What marks it out from other fields of human endeavor is that, because of its formalized humility, it's always ready to correct itself when it makes a mistake. Scientists may not be humble people, but doing science forces us to act humbly."

Carl Sagan saw it differently: "In science it often happens that scientists say, 'You know that's a really good argument; my position is mistaken', and then they would actually change their minds and you never hear that old view from them again. They really do it. It doesn't happen as often as it should, because scientists are human and change is sometimes painful. But it happens every day. I cannot recall the last time something like that happened in politics or religion." Carl Sagan was wrong about religion. He forgot Christianity, Reformation, and the origin of universities. He was even more wrong about scientists changing their minds every day.

To refute a false claim is for the benefit of all. To reject the practice of repeating the false claim, and to name those who keep repeating claims that are known to be untrue, is not defamation of such authors. On the contrary, it is a service to all who use science, which include the authors of erroneous claims. This is why authors and journals publish errata and retractions, and why universities worth their name should punish those who plagiarize and publish fake science. The relentless pursuit of the truth is in the public interest.

Science is self-correcting because it is imbibed in freedom. This key truth of science needs to be broadly communicated to all, not just scientists.

References

1. A. Bejan, Thermodynamics today. Energy **160**, 1208–1219 (2018)
2. A. Bejan, Evolution in thermodynamics. Appl. Phys. Rev. **4** (2017), article 011305
3. A. Bejan, Thermodynamics of heating. Prof. R. Soc. A **475** (2019), article 20180820
4. A. Bejan, *Advanced Engineering Thermodynamics*, 4th edn. (Wiley, Hoboken, 2016)
5. L.E. Dugatkin, L. Trut, *How to Tame a Fox* (The University of Chicago Press, Chicago, 2017)
6. A. Bejan, Comment on "Study on the consistency between field synergy principle and entransy dissipation extremum principle". Int. J. Heat Mass Transf. **120**, 1187–1188 (2018)
7. A. Bejan, Letter to the Editor on "Temperature-heat diagram analysis method for heat recovery physical adsorption refrigeration cycle—Taking multi stage cycle as an example. Int. J. Refrig **90**, 277–279 (2018)
8. A. Bejan, S. Lorente, Letter to the editor. Chem. Eng. Process. **56**, 34 (2012)
9. A. Bejan, "Entransy", and its lack of content in physics. J. Heat Trans. **136**, 055501 (2014)
10. A. Bejan, Comment on "Application of entransy analysis in self-heat recuperation technology". Ind. Eng. Chem. Res. **53**, 1274–1285 (2014)
11. A. Bejan, Heatlines (1983) versus synergy, Int. J. Heat Mass Trans. **81**, 2015, 654–658 (1998)
12. A. Bejan, Letter to the editor of renewable and sustainable energy reviews. Renew. Sustain. Energy Rev. **53**, 1636–1637 (2016)
13. B.-C. Han, *Deconstruction in Chinese* (MIT Press, Cambridge, MA, 2017)
14. A. Qin, Fraud scandals sap China's dream of becoming a science superpower. The New York Times, October 13, 2017
15. W. Quang, B. Chen, F. Shu, Publish or impoverish: an investigation of the monetary reward system of science in China (1999–2016). Aslib J. Inf. Manag. **69**, 486–502 (2017)
16. A. Abritis, A. McCook, Cash incentives for papers go global. Science **357**, 541 (2017)
17. D.A. Eisner, Reproducibility of science: Fraud, impact factors and carelessness. J. Mol. Cellular Cardiol., 21 October 2017, https://doi.org/10.1016/j.yjmcc.2017.10.009
18. I. Fister Jr, I. Fister and M. Perc, Toward the discovery of citation cartels in citation networks. Front. Phys. 4 (2016), Article 49
19. A. Bejan, Plagiarism is not a victimless crime, *Prism*. Am. Soc. Eng. Educ. **28**(7), 52 (2019)
20. E. Chiscop-Head, Research integrity interview series: If you cheat, there should be a referee who blows the whistle against you, Duke University School of Medicine, 20 September 2019
21. M.P. Taylor, *Science is enforced humility*. The Guardian, 13 November 2012

Index

A. Bejan, *Freedom and Evolution*,
https://doi.org/10.1007/978-3-030-34009-4

Printed in the United States
By Bookmasters